高等职业教育"十三五"规划教材（移动互联应用技术专业）

Android 项目开发实战

主　编　赵善龙　李旭东

副主编　姜培育　李春宝

中国水利水电出版社
www.waterpub.com.cn
·北京·

内 容 提 要

本书规划了 Android 从入门到进阶过程中最重要的知识体系，将知识、技术与技巧充分渗透到多个独立且完整的 Android 应用实战项目中，带领读者一同参与到真正的企业开发流程，使读者有条不紊地掌握完整的项目开发技术，循序渐进地具备企业级移动应用开发的能力。在程序实例讲解方面，注重对实际动手能力的指导；在遵循项目开发过程的同时，将重要知识点和经验技巧以"关键知识点解析"的形式呈现给读者，这为初学者将学习与实践结合提供了很好的指导。

本书可作为大学本科和高职高专有关课程的实训教材，也可供具备一定手机开发经验的开发者及 Android 开发爱好者参考和使用。

本书提供实例源代码，读者可以从万水书苑以及中国水利水电出版社网站下载，网址为 http://www.wsbookshow.com 和 http://www.waterpub.com.cn/softdown/。

图书在版编目（CIP）数据

Android项目开发实战 / 赵善龙, 李旭东主编. --北京：中国水利水电出版社, 2018.8
高等职业教育"十三五"规划教材. 移动互联应用技术专业
ISBN 978-7-5170-6702-3

Ⅰ. ①A… Ⅱ. ①赵… ②李… Ⅲ. ①移动终端－应用程序－程序设计－高等职业教育－教材 Ⅳ. ①TN929.53

中国版本图书馆CIP数据核字(2018)第175367号

策划编辑：周益丹　责任编辑：张玉玲　加工编辑：张溯源　封面设计：李　佳

书　名	高等职业教育"十三五"规划教材（移动互联应用技术专业） **Android 项目开发实战** Android XIANGMU KAIFA SHIZHAN
作　者	主　编　赵善龙　李旭东 副主编　姜培育　李春宝
出版发行	中国水利水电出版社 （北京市海淀区玉渊潭南路 1 号 D 座　100038） 网址：www.waterpub.com.cn E-mail: mchannel@263.net（万水） 　　　　sales@waterpub.com.cn 电话：（010）68367658（营销中心）、82562819（万水）
经　售	全国各地新华书店和相关出版物销售网点
排　版	北京万水电子信息有限公司
印　刷	三河市铭浩彩色印装有限公司
规　格	184mm×260mm　16 开本　17 印张　420 千字
版　次	2018 年 8 月第 1 版　2018 年 8 月第 1 次印刷
印　数	0001—3000 册
定　价	36.00 元

凡购买我社图书，如有缺页、倒页、脱页的，本社营销中心负责调换
版权所有·侵权必究

前　　言

在移动编程技术中，Android 将开发者使用最多的 Java 语言作为基础语言，为众多已掌握 Java 编程技术的开发者降低了学习门槛。同时，谷歌对 Android 进行了从组件到 UI 各层次较完善和丰富的封装，为开发者提供了大量简洁易用的 API 和基础 UI 控件，这也在很大程度上降低了开发者的学习成本。然而摆在众多开发者面前的问题是，很多开发者了解 Android 中 Activity 的生命周期各环节被调用的时机，但却对生命周期各环节应该编写哪部分代码不得其法；很多开发者了解 UI 控件的使用方法，但却在通过网络获取数据后刷新 UI 方面力不从心；很多开发者了解如何使用 ListView 展示列表数据，但当数据量稍微增大时程序就会出现卡顿甚至崩溃；很多开发者可以熟练地绘制布局并在模拟器上显示完整，但一旦到了某些真机上，画面却惨不忍睹。事实上，如何综合地运用 Android 开发技术进行规范的应用开发，如何使自己的开发技术与企业开发流程接轨，如何更好地优化应用，使应用适配更广泛的机型而且程序更加健壮，的确是使许多通过自学成长的开发者深受困扰的问题。凭借多年的院校教学经验和企业实践经验，我们深知 Android 初学者在学习和成长过程中的痛点。针对这些痛点，本书规划了 Android 从入门到进阶过程中最重要的知识体系，将知识、技术与技巧充分渗透到多个独立且完整的 Android 应用实战项目中，带领读者一同参与到真正的企业开发流程中，使读者有条不紊地掌握完整的项目开发技术，循序渐进地具备企业级移动应用开发的能力。

在开始项目实战之前，需要读者对本书的知识结构体系图进行初步了解，读者应在掌握预备知识的基础上对本书项目进行逐一学习。本书将着重对基础组件、UI、线程与线程间通信、网络通信、数据解析和数据存储六大部分在项目中的应用进行讲解。下面介绍每个项目重点训练的知识点。

项目 1 主要针对项目构建、布局、基础控件和按钮的点击事件进行实战。

项目 2 主要针对应用的架构搭建、ListView 的使用方法和技巧、自定义 Adapter 的方法进行实战。

项目 3 主要针对本地文件存储、自定义控件、onTouch 事件处理及 Canvas 进行实战。

项目 4 主要针对图片处理及优化、文件读写、ContentProvider 进行实战。

项目 5 主要针对线程间通信、时钟、Service 进行实战。

项目 6 主要针对线程间通信、HTTP、BroadcastReceiver、Service 进行实战。

项目 7 主要针对 Fragment、XML 解析、WebView 进行实战。

项目 8 主要针对 HttpClient、JSON 解析、网络通信的封装进行实战。

项目 9 主要针对 BroadcastReceiver、Service、AIDL 和电话操作进行实战。

项目 10 主要针对蓝牙通信进行实战。

项目 11 主要针对 Socket、Handle 和消息队列进行实战。

项目 12 主要针对百度地图、定位、SQLiteOpenHelper 和自动更新进行实战。

本书由赵善龙、李旭东任主编，姜培育、李春宝任副主编，另外还要感谢周益丹编辑对本书提出了非常宝贵的意见，特别是书中内容的编排、难易程度的把握、案例的选取和文叙风格等。

由于编者水平有限，书中不妥之处在所难免，恳请读者批评指正。

编　者
2018 年 6 月

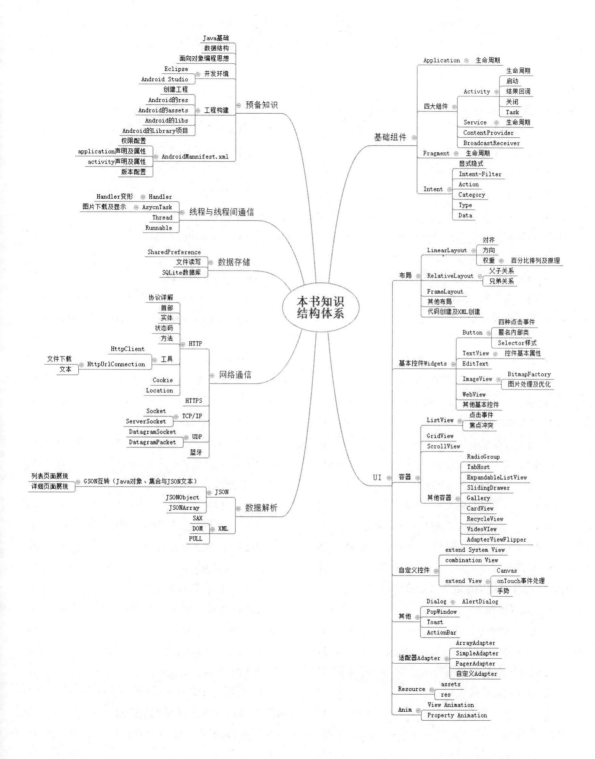

目　　录

前言

项目1　简易计算器 ·· 1
  1.1　总体设计 ·· 1
    1.1.1　总体分析 ·· 1
    1.1.2　功能模块框图 ·································· 2
    1.1.3　系统流程图 ···································· 2
    1.1.4　界面设计 ·· 2
  1.2　详细设计 ·· 4
    1.2.1　模块描述 ·· 4
    1.2.2　系统包及其资源规划 ·························· 5
    1.2.3　主要方法流程设计 ···························· 5
  1.3　代码实现 ·· 7
    1.3.1　显示界面布局 ·································· 7
    1.3.2　控件设计实现 ·································· 7
    1.3.3　控件事件处理方法实现 ····················· 12
    1.3.4　数值计算方法实现 ·························· 15
  1.4　关键知识点解析 ··································· 16
    1.4.1　在程序中创建菜单 ·························· 16
    1.4.2　基础界面布局 ································ 19
    1.4.3　设置程序名称和图标 ························ 21
    1.4.4　常用文本输入控件及按钮 ·················· 21
    1.4.5　为按钮增加多种样式——selector ········ 26
    1.4.6　多分辨率适配利器——LinearLayout · 28
  1.5　问题与讨论 ··· 30

项目2　基于离线数据的天气应用——
      天气预报（一） ··································· 31
  2.1　总体设计 ·· 31
    2.1.1　总体分析 ······································ 31
    2.1.2　功能模块框图 ································ 32
    2.1.3　系统流程图 ··································· 32
    2.1.4　界面设计 ······································ 33
  2.2　详细设计 ·· 34
    2.2.1　模块描述 ······································ 34
    2.2.2　系统包及其资源规划 ······················· 36

    2.2.3　主要方法流程设计 ·························· 39
  2.3　代码实现 ·· 39
    2.3.1　显示界面布局 ································ 39
    2.3.2　控件设计实现 ································ 39
    2.3.3　主要代码功能分析 ·························· 43
  2.4　关键知识点解析 ··································· 49
    2.4.1　ListView控件的用法 ······················· 49
    2.4.2　自定义适配器 ································ 52
    2.4.3　GridView控件的用法 ····················· 57
    2.4.4　ViewPager控件的用法 ··················· 57
    2.4.5　ListView中存在按钮时导致ListItem
           点击无效的解决方案 ·························· 60
  2.5　问题与讨论 ··· 60

项目3　天天爱读书手机阅读器 ······················ 61
  3.1　总体设计 ·· 61
    3.1.1　功能模块框图 ································ 62
    3.1.2　系统流程图 ··································· 62
    3.1.3　界面设计 ······································ 63
  3.2　详细设计 ·· 63
    3.2.1　模块描述 ······································ 63
    3.2.2　系统包及其资源规划 ······················· 64
    3.2.3　主要方法流程设计 ·························· 66
  3.3　代码实现 ·· 67
    3.3.1　显示界面布局 ································ 67
    3.3.2　Touch事件方法实现 ······················· 68
  3.4　关键知识点解析 ··································· 69
    3.4.1　Android的Touch事件处理机制 ······· 69
    3.4.2　掌握自定义控件 ····························· 70
  3.5　问题与讨论 ··· 73

项目4　基于本地图库的图片应用——
      幻彩手机相册 ······································ 74
  4.1　总体设计 ·· 74
    4.1.1　功能模块框图 ································ 74

    4.1.2　系统流程图……………………74
    4.1.3　界面设计………………………76
  4.2　详细设计……………………………76
    4.2.1　模块描述………………………76
    4.2.2　系统包及其资源规划…………77
    4.2.3　主要方法流程设计……………80
  4.3　代码实现……………………………81
    4.3.1　显示界面布局…………………81
    4.3.2　读取手机图库方法实现………82
    4.3.3　图片方向的判断………………83
    4.3.4　图片压缩………………………84
    4.3.5　使用 Android 提供的媒体播放器
          （MediaPlayer）…………………85
  4.4　关键知识点解析……………………87
    4.4.1　图片加载到内存 OOM…………87
    4.4.2　大量图片的缓存处理…………89
  4.5　问题与讨论…………………………90

项目 5　学习监督器……………………91
  5.1　总体设计……………………………91
    5.1.1　总体分析………………………91
    5.1.2　功能模块框图…………………91
    5.1.3　系统流程图……………………92
    5.1.4　界面设计………………………92
  5.2　详细设计……………………………94
    5.2.1　系统包及其资源规划…………94
    5.2.2　时间设置 Activity 设计………94
    5.2.3　后台服务设计…………………95
  5.3　代码实现……………………………96
    5.3.1　显示界面布局…………………96
    5.3.2　构建一个服务…………………97
    5.3.3　创建启动服务…………………99
    5.3.4　监控网络变化…………………102
    5.3.5　时间比较………………………103
    5.3.6　创建绑定的服务………………104
    5.3.7　使用 Activity 作为 Dialog……105
  5.4　关键知识点解析……………………106
    5.4.1　在前台运行服务………………106
    5.4.2　服务的生命周期………………106
    5.4.3　避免系统回收服务……………107

  5.5　问题与讨论…………………………108

项目 6　简易网络音乐播放器…………109
  6.1　总体设计……………………………109
    6.1.1　总体分析………………………109
    6.1.2　功能模块框图…………………110
    6.1.3　系统流程图……………………110
    6.1.4　界面设计………………………110
  6.2　详细设计……………………………111
    6.2.1　模块描述………………………111
    6.2.2　系统包及其资源规划…………113
    6.2.3　主要方法流程设计……………115
  6.3　代码实现……………………………115
    6.3.1　显示界面布局…………………115
    6.3.2　HttpURLConnection 网络通信方法
          实现…………………………116
    6.3.3　XML 数据解析方法实现………117
  6.4　关键知识点解析……………………118
    6.4.1　AsyncTask（异步任务）的使用……118
    6.4.2　HttpClient、HttpURLConnection、
          okHttp 和 Volley 的网络通信对比……119
    6.4.3　HttpClient 和 HttpURLConnection 的
          使用方法……………………120
  6.5　问题与讨论…………………………122

项目 7　新闻客户端……………………123
  7.1　总体设计……………………………123
    7.1.1　功能模块框图…………………123
    7.1.2　系统流程图……………………124
    7.1.3　界面设计………………………124
  7.2　详细设计……………………………125
    7.2.1　模块描述………………………125
    7.2.2　系统包及其资源规划…………126
    7.2.3　主要方法流程设计……………128
  7.3　代码实现……………………………129
    7.3.1　显示界面布局…………………129
    7.3.2　RSS 内容读取方法实现………130
    7.3.3　利用 WebView 显示 HTML 页面……133
    7.3.4　利用 ViewHolder 优化 AdapterView·133
    7.3.5　Fragment 的简单使用方法
          （FragmentStatePagerAdapter）……134

7.3.6 菜单的使用技巧（ActionBar）……… 135
7.4 关键知识点解析……………………… 135
　7.4.1 用户体验………………………… 135
　7.4.2 RSS 阅读器实现………………… 136
　7.4.3 深入理解 XML 数据格式………… 139
7.5 问题与讨论…………………………… 145

项目 8 基于网络通信的天气应用——天气预报（二）……………………… 146
8.1 总体设计……………………………… 146
　8.1.1 总体分析………………………… 146
　8.1.2 功能模块框图…………………… 147
　8.1.3 系统流程图……………………… 147
　8.1.4 界面设计………………………… 147
8.2 详细设计……………………………… 148
　8.2.1 模块描述………………………… 148
　8.2.2 系统包及其资源规划…………… 150
　8.2.3 主要方法流程设计……………… 152
8.3 代码实现……………………………… 153
　8.3.1 显示界面布局…………………… 153
　8.3.2 控件设计实现…………………… 153
　8.3.3 天气预报接口方法实现………… 153
8.4 关键知识点解析……………………… 155
　8.4.1 在程序中使用天气预报接口…… 155
　8.4.2 采用 MQTT 协议实现 Android
　　　　推送…………………………… 158
8.5 问题与讨论…………………………… 167

项目 9 商务通讯录………………………… 168
9.1 总体设计……………………………… 168
　9.1.1 总体分析………………………… 168
　9.1.2 功能模块框图…………………… 168
　9.1.3 系统流程图……………………… 169
　9.1.4 界面设计………………………… 169
9.2 详细设计……………………………… 170
　9.2.1 模块描述………………………… 170
　9.2.2 系统包及其资源规划…………… 172
　9.2.3 主要方法流程设计……………… 173
9.3 代码实现……………………………… 175
　9.3.1 显示界面布局…………………… 175
　9.3.2 控件设计实现…………………… 176

　9.3.3 监听手机来电服务……………… 179
　9.3.4 挂断电话………………………… 180
9.4 关键知识点解析……………………… 182
　9.4.1 进程通信——AIDL 的使用…… 182
　9.4.2 双卡双待手机如何获取来电…… 186
9.5 问题与讨论…………………………… 187

项目 10 蓝牙打印机……………………… 188
10.1 总体设计…………………………… 188
　10.1.1 总体分析……………………… 188
　10.1.2 功能模块框图………………… 188
　10.1.3 系统流程图…………………… 189
　10.1.4 界面设计……………………… 189
10.2 详细设计…………………………… 190
　10.2.1 模块描述……………………… 190
　10.2.2 系统包及其资源规划………… 192
　10.2.3 主要方法流程设计…………… 194
10.3 代码实现…………………………… 195
　10.3.1 显示界面布局………………… 195
　10.3.2 控件设计实现………………… 197
　10.3.3 获取图片分享………………… 201
　10.3.4 蓝牙设备和设置可见时间…… 203
　10.3.5 搜索蓝牙设备………………… 203
　10.3.6 连接蓝牙设备………………… 204
　10.3.7 蓝牙通信……………………… 205
10.4 关键知识点解析…………………… 206
　10.4.1 静默开启蓝牙………………… 206
　10.4.2 蓝牙自动配对………………… 207
10.5 问题与讨论………………………… 209

项目 11 基于 Socket 的 Bmop 即时通信……… 210
11.1 总体设计…………………………… 210
　11.1.1 总体分析……………………… 210
　11.1.2 功能模块框图………………… 210
　11.1.3 系统流程图…………………… 211
　11.1.4 界面设计……………………… 211
11.2 详细设计…………………………… 212
　11.2.1 模块描述……………………… 212
　11.2.2 系统包及其资源规划………… 213
　11.2.3 主要方法流程设计…………… 216
11.3 代码实现…………………………… 217

11.3.1	显示界面布局	217
11.3.2	控件设计实现	219
11.3.3	Socket 线程	227
11.3.4	待发消息队列	228
11.3.5	消息接收	229

11.4 关键知识点解析 233
    11.4.1 Socket 定义 233
    11.4.2 Socket 与 HTTP 对比 233
    11.4.3 使用 UDP 协议通信 233

11.5 问题与讨论 234

项目 12 易行打车 235

12.1 总体设计 235
    12.1.1 总体分析 235
    12.1.2 功能模块框图 235
    12.1.3 系统流程图 236
    12.1.4 界面设计 236

12.2 详细设计 237
    12.2.1 模块描述 237
    12.2.2 系统包及其资源规划 238
    12.2.3 主要方法流程设计 241

12.3 代码实现 242
    12.3.1 显示界面布局 242
    12.3.2 控件设计实现 244
    12.3.3 申请百度地图 API Key 250
    12.3.4 初始化定位 251
    12.3.5 定位监听 251
    12.3.6 初始化地图 View 252
    12.3.7 显示位置信息 252
    12.3.8 获取当前屏幕的经纬度范围 253
    12.3.9 增加多个标注并监听 253

12.4 关键知识点解析 255
    12.4.1 在线更新 255
    12.4.2 Android 的四种定位方式 260

12.5 问题与讨论 262

项目 1　简易计算器

本项目从简易计算器入手,为读者呈现一套较完整的简易计算器项目的建设流程,模拟企业级原生移动应用开发的主要环节,从项目总体分析、功能模块拆分、操作流程分析、功能及界面设计、编码等多个重要环节对项目进行讲述。项目虽小,却能够让读者初步体验到企业级移动应用开发的基本方法。

项目需求描述如下:

(1)要求使用 Android 原生开发技术实现一款具有加、减、乘、除运算功能的计算器应用。

(2)用户可以按数学运算法则输入数字和运算符号进行运算,运算结果可以参与下一次运算,运算结果支持 12 位显示。

(3)支持正值、负值、小数的运算。

(4)支持非法输入的验证及提示。

(5)支持退格以修改数据。

(6)支持清屏和重置。

(7)支持展示最近几次的计算记录。

(1)掌握基础界面布局。

(2)掌握输入控件、按钮控件。

(3)控件 OnClick 事件的多种实现方式。

(4)使用百分比技巧进行布局。

1.1　总体设计

1.1.1　总体分析

根据项目需求进行分析,简易计算器应实现以下功能:

(1)拥有友好的计算器界面,便于用户使用。

(2)显示 12 位结果,具有基本的加、减、乘、除功能。

(3)能够判断用户输入运算数是否正确。

(4)支持小数运算。

(5)具有退格功能,能够删除最后一个输入。

（6）具有清除功能。

（7）具有结果存储功能，能够显示存储器状态，支持触屏手机。

整个程序除总体模块外，主要分为输入模块、显示模块以及计算模块（包括一些其他功能）三大部分。

总体模块控制系统的生命周期，输入模块部分负责读取用户输入的数据，显示模块部分负责显示用户之前输入的数据以及显示最终的计算结果，计算模块部分负责进行数据的运算以及一些辅助的功能。

1.1.2 功能模块框图

根据总体分析结果可以总结出计算器功能模块框图，如图 1-1 所示。

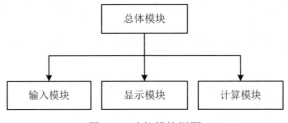

图 1-1 功能模块框图

本计算器主要可以实现基本的加减乘除运算和其他一些运算。基本运算包括加法运算、减法运算、乘法运算和除法运算，其他运算包括开方运算、正负运算、清除运算等，如图 1-2 所示。

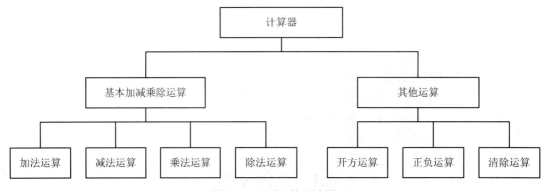

图 1-2 主要运算方法图

1.1.3 系统流程图

根据总体分析结果及功能模块框图梳理出系统主要流程，如图 1-3 所示。

1.1.4 界面设计

在系统总体分析及功能模块划分清楚后，就要考虑界面的设计了。界面设计应该尽量简洁而美观，应该具有良好的交互性。

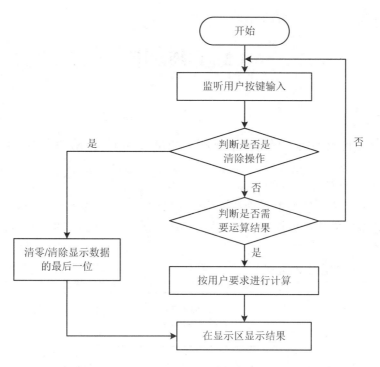

图 1-3　系统流程图

主界面是程序操作过程中必不可少的，它是与用户交互过程中的重要环节。通过主界面，用户可以调用系统相关的各个模块，使用系统中实现的各个功能。计算器界面如图 1-4 所示。

图 1-4　计算器界面设计图

从图 1-4 中我们可以很直观地看到，计算器界面区域主要有数据和结果显示区、数字按键区和计算按键区。

数据和结果显示区用于显示用户输入的数据、最终的计算结果和一些其他信息。

数字按键区和计算按键区主要布置计算器键盘，在触屏上提供各种功能的按键，负责读取用户的键盘输入以及响应触屏的按键。

1.2 详细设计

1.2.1 模块描述

在系统总体分析及界面布局设计完成后,主要工作就转入对各个功能模块的详细设计阶段。

1. 总体模块详细设计

总体模块需要完成的任务主要是系统的程序启动类,它需要负责整个系统的生命周期。同时还要在该模块中完成菜单栏的所有功能,即退出程序、记忆数据、显示数据和清除记忆数据这四个功能。

总体模块功能图如图 1-5 所示。

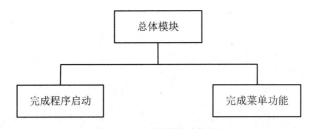

图 1-5　总体模块功能图

2. 输入模块详细设计

输入模块的主要任务是描述计算器键盘以及实现键盘的监听,即当用户点击按键或者屏幕的时候,监听器会去调用相应的处理办法或其他相应的处理模块。本模块还为系统提供一个较为直观的键盘图形用户界面。

输入模块功能图如图 1-6 所示。

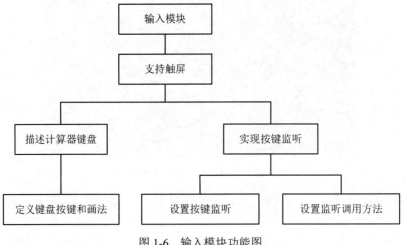

图 1-6　输入模块功能图

3. 计算模块详细设计

系统要完成整个计算器的计算功能，计算模块就是整个系统的重点模块。没有计算模块系统就不能顺利地完成计算，也就无法达到用户的要求。所以计算模块的设计也是本次系统设计中的重点。

当输入模块的监听传到计算模块中时，计算模块就要根据相应的方法进行进一步处理。

计算模块功能图如图 1-7 所示。

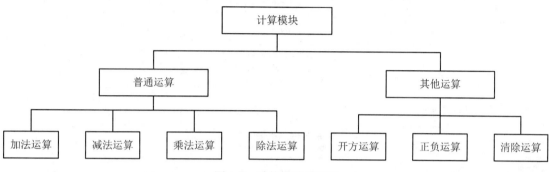

图 1-7　计算模块功能图

4. 显示模块详细设计

显示模块需要对计算器的计算区域进行显示，使用户能够看到整个计算器的画面。该区域的显示信息包括用户输入的数据、最终的计算结果和一些其他信息。同时本模块还将提供调用和设置显示的具体方法。

显示模块功能图如图 1-8 所示。

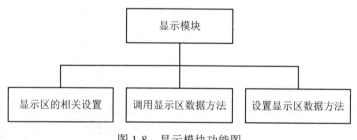

图 1-8　显示模块功能图

1.2.2　系统包及其资源规划

在系统各个模块的实现方式和流程设计完成后，我们将对系统主要的包和资源进行规划，划分的原则主要是保持各个包相互独立，耦合度尽量低。

根据系统功能设计，本系统仅需一个 Activity 类，系统的几个功能实现方式基本相同，因此可以按照一个包规划，在包中设计不同的方法支持不同的功能。包及其资源结构如图 1-9 所示。

1.2.3　主要方法流程设计

根据之前的分析和功能划分情况梳理出系统的主要方法流程，如图 1-10 所示。

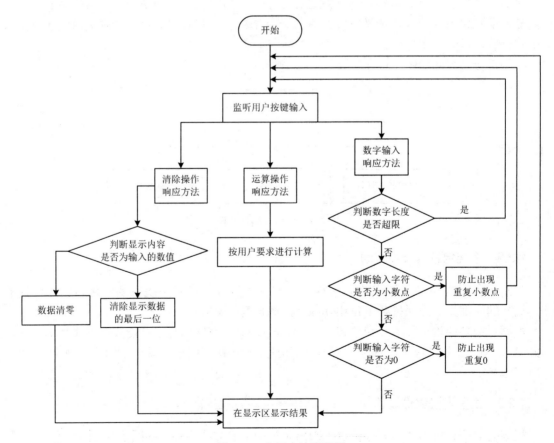

图 1-9　包及其资源结构

图 1-10　主要方法流程图

1.3 代码实现

1.3.1 显示界面布局

系统主界面是进入系统后显示的界面，该界面包括一个 ScrollView、两个 TextView 和若干 Button，如图 1-11 所示。

图 1-11 系统主界面

1.3.2 控件设计实现

在工程的 res/layout 目录下，创建一个名称为 activity_main.xml 的布局文件（部分代码如下所示），其外层是 RelativeLayout，内层嵌套 LinearLayout 以完成布局管理。该布局文件中有一个 ScrollView，其中放置了一个 TextView，用于显示计算记录，另一个 TextView 用于显示输入数值和计算结果，19 个 Button 用于用户输入和计算操作控制。我们可以看到，LinearLayout 中使用了 android:layout_weight 属性，这个属性使 LinearLayout 具有按百分比对子控件进行布局的能力。我们将在 1.4.6 节中对 LinearLayout 这种特性进行详细讲解。

```
<RelativeLayout xmlns:android="http://schemas.android.com/apk/res/android"
    xmlns:tools="http://schemas.android.com/tools"
    android:layout_width="match_parent"
    android:layout_height="match_parent"
    android:paddingBottom="@dimen/activity_vertical_margin"
    android:paddingLeft="@dimen/activity_horizontal_margin"
    android:paddingRight="@dimen/activity_horizontal_margin"
    android:paddingTop="@dimen/activity_vertical_margin"
    tools:context="edu.hrbeu.kmq.calculator.MainActivity" >

    <LinearLayout
        android:layout_width="match_parent"
        android:layout_height="match_parent"
```

```xml
        android:layout_centerHorizontal="true"
        android:layout_centerVertical="true"
        android:gravity="bottom"
        android:orientation="vertical" >

        <ScrollView
            android:id="@+id/scrollView1"
            android:layout_width="match_parent"
            android:layout_height="0dp"
            android:layout_weight="1"
            android:scrollbars="vertical" >

            <LinearLayout
                android:layout_width="match_parent"
                android:layout_height="wrap_content"
                android:orientation="vertical" >

                <TextView
                    android:id="@+id/textViewCalcHistory"
                    android:layout_width="fill_parent"
                    android:layout_height="wrap_content"
                    android:gravity="left|bottom"
                    android:text="计算记录"
                    android:typeface="monospace" />
            </LinearLayout>
        </ScrollView>

        <TextView
            android:id="@+id/textViewDisplayArea"
            android:layout_width="match_parent"
            android:layout_height="wrap_content"
            android:layout_marginBottom="10dp"
            android:gravity="right|center_vertical"
            android:text="Display Area"
            android:textAppearance="?android:attr/textAppearanceLarge"
            android:textSize="28sp"
            android:typeface="monospace" />

        <LinearLayout
            android:layout_width="match_parent"
            android:layout_height="wrap_content" >

            <Button
                android:id="@+id/Button7"
                android:layout_height="wrap_content"
                android:layout_width="0dp"
```

```xml
            android:layout_weight="1"
            android:text="7"
            android:onClick="addNumber" />

        <Button
            android:id="@+id/Button8"
            android:layout_height="wrap_content"
            android:layout_width="0dp"
            android:layout_weight="1"
            android:text="8"
            android:onClick="addNumber" />

        <Button
            android:id="@+id/Button9"
            android:layout_height="wrap_content"
            android:layout_width="0dp"
            android:layout_weight="1"
            android:text="9"
            android:onClick="addNumber" />

        <Button
            android:id="@+id/ButtonMultiply"
            android:layout_height="wrap_content"
            android:layout_width="0dp"
            android:layout_weight="1"
            android:text="×"
            android:onClick="doCalc" />

        <Button
            android:id="@+id/ButtonDivide"
            android:layout_height="wrap_content"
            android:layout_width="0dp"
            android:layout_weight="1"
            android:text="÷"
            android:onClick="doCalc" />
</LinearLayout>

<LinearLayout
    android:layout_width="match_parent"
    android:layout_height="wrap_content" >

        <Button
            android:id="@+id/Button4"
            android:layout_height="wrap_content"
            android:layout_width="0dp"
            android:layout_weight="1"
```

```xml
            android:text="4"
            android:onClick="addNumber" />

        <Button
            android:id="@+id/Button5"
            android:layout_height="wrap_content"
            android:layout_width="0dp"
            android:layout_weight="1"
            android:text="5"
            android:onClick="addNumber" />

        <Button
            android:id="@+id/Button6"
            android:layout_height="wrap_content"
            android:layout_width="0dp"
            android:layout_weight="1"
            android:text="6"
            android:onClick="addNumber" />

        <Button
            android:id="@+id/ButtonPlus"
            android:layout_height="wrap_content"
            android:layout_width="0dp"
            android:layout_weight="1"
            android:text="+"
            android:onClick="doCalc" />

        <Button
            android:id="@+id/ButtonMinus"
            android:layout_height="wrap_content"
            android:layout_width="0dp"
            android:layout_weight="1"
            android:text="-"
            android:onClick="doCalc" />
    </LinearLayout>

    <LinearLayout
        android:layout_width="match_parent"
        android:layout_height="wrap_content" >

        <Button
            android:id="@+id/Button1"
            android:layout_height="wrap_content"
            android:layout_width="0dp"
            android:layout_weight="1"
            android:text="1"
```

```xml
            android:onClick="addNumber" />

        <Button
            android:id="@+id/Button2"
            android:layout_height="wrap_content"
            android:layout_width="0dp"
            android:layout_weight="1"
            android:text="2"
            android:onClick="addNumber" />

        <Button
            android:id="@+id/Button3"
            android:layout_height="wrap_content"
            android:layout_width="0dp"
            android:layout_weight="1"
            android:text="3"
            android:onClick="addNumber" />

        <Button
            android:id="@+id/ButtonSqrt"
            android:layout_height="wrap_content"
            android:layout_width="0dp"
            android:layout_weight="1"
            android:text="√"
            android:onClick="calcSqrt" />

        <Button
            android:id="@+id/ButtonDel"
            android:layout_height="wrap_content"
            android:layout_width="0dp"
            android:layout_weight="1"
            android:text="←"
            android:onClick="delNumber"/>

</LinearLayout>

<LinearLayout
    android:layout_width="match_parent"
    android:layout_height="wrap_content" >

        <Button
            android:id="@+id/Button0"
            android:layout_height="wrap_content"
            android:layout_width="0dp"
            android:layout_weight="1"
            android:text="0"
```

```xml
            android:onClick="addNumber" />

        <Button
            android:id="@+id/ButtonDot"
            android:layout_height="wrap_content"
            android:layout_width="0dp"
            android:layout_weight="1"
            android:text="."
            android:onClick="addNumber" />

        <Button
            android:id="@+id/ButtonPosNeg"
            android:layout_height="wrap_content"
            android:layout_width="0dp"
            android:layout_weight="1"
            android:text="+/-"
            android:onClick="toggleSign" />

        <Button
            android:id="@+id/ButtonCalc"
            android:layout_height="wrap_content"
            android:layout_width="0dp"
            android:layout_weight="2"
            android:text="="
            android:onClick="doCalc" />
        </LinearLayout>
    </LinearLayout>
</RelativeLayout>
```

1.3.3 控件事件处理方法实现

1. MainActivity 的创建

创建一个名称为 MainActivity 的 Activity，在类的顶部声明用到的 ScrollView、TextView 和 Button 组件，在 onCreate()方法中调用 setContentView(R.layout.activity_main)方法设置布局视图，通过 findViewById()方法实例化 ScrollView、TextView 和 Button 对象。

主要代码如下：

```java
package edu.hrbeu.kmq.calculator;
import android.app.Activity;
import android.app.AlertDialog;
import android.app.AlertDialog.Builder;
import android.text.ClipboardManager;
import android.content.Context;
import android.content.DialogInterface;
import android.content.DialogInterface.OnClickListener;
import android.content.Intent;
import android.os.Bundle;
```

```java
import android.view.Menu;
import android.view.MenuItem;
import android.view.View;
import android.widget.Button;
import android.widget.ScrollView;
import android.widget.TextView;
public class MainActivity extends    Activity {

    double nowNumber,prevNumber;                    //当前显示数值（double格式）、上一个操作数
    string nowString;                               //当前显示数值（string格式）
    char savedOp;                                   //保存的运算符
    boolean isUserInput,isDotInputed,hasPrevNumber; //当前数字是用户输入的、当前已输入小数点、
                                                    //有上一个操作数

    @Override
    protected void onCreate(Bundle savedInstanceState) {
        super.onCreate(savedInstanceState);
        setContentView(R.layout.activity_main);
        allClear(null);
        ((Button) findViewById(R.id.ButtonDel)).setOnLongClickListener(
            new View.OnLongClickListener() {
                @Override
                public boolean onLongClick(View v) {
                    allClear(null);
                    return true;
                }
            }
        );
    }
    …
    …
}
```

2. 数值输入响应方法 addNumber()

在输入数值时，需要判断当前显示的是什么数值。如果当前显示的是正常已经输入的数值，则将最新输入的数值添加到原有数值的后面；如果当前显示的是原有的运算结果，则先清除原有数据；如果只是中间结果，则在清除原有数据时保留原有的中间结果。

在输入数值时，对于 0 和小数点需要判断重复输入的情况。主要代码如下：

```java
public void addNumber(View view) {
    Button button = (Button) view;
    char newChar = ((String) button.getText()).charAt(0);
    if (!isUserInput){                          //如果是运算结果，则先清除所有数据
        boolean hasPrev=hasPrevNumber;          //如果只是中间结果，要保留中间状态
        allClear(null);
        hasPrevNumber=hasPrev;
    }
```

```
            if (isDotInputed) {                    //最多限制到12位（不含小数点）
                if (nowString.length() >= 13)
                    return;
            } else {
                if (nowString.length() >= 12)
                    return;
            }
            switch (newChar) {
                case '0':                           //避免出现输入多个0的情况
                    if (nowString == "0")           //比如000000
                        return;
                    break;
                case '.':
                    if (isDotInputed)               //如果已经输入了小数点就不能再输入
                        return;
                    else
                        isDotInputed = true;
                    break;
                default:
                    if (nowString == "0")           //避免前导0
                        nowString = "";
            }
            nowString += newChar;
            refreshAndSync();
        }
```

3. 运算符输入响应方法 doCalc ()

在输入运算符时，需要判断之前是否已输入过运算符。如果之前有过输入，则先进行一次计算再记录本次输入的运算符；如果之前没有过输入，则直接记录本次输入的运算符。主要代码如下：

```
        public void doCalc(View view) {
            isUserInput = false;
            if (hasPrevNumber) {
                calc();
                refreshAndSync();
            }
            hasPrevNumber = true;
            prevNumber = nowNumber;
            switch (view.getId()) {
            case R.id.ButtonPlus:
                savedOp = '+';
                break;
            case R.id.ButtonMinus:
                savedOp = '-';
                break;
            case R.id.ButtonMultiply:
                savedOp = '*';
```

```
            break;
        case R.id.ButtonDivide:
            savedOp = '/';
            break;
        case R.id.ButtonCalc:
        default:
            savedOp = '=';
            hasPrevNumber = false;
            break;
    }
}
```

4. 清除操作符输入响应方法 delNumber ()

在输入清除操作符时，需要判断当前清除内容是用户输入的数值还是运算结果。如果是用户输入的数值，则删除最后一个数字符；如果是运算结果，则将运算结果的内容全部清零。主要代码如下：

```
public void delNumber(View view) {
    if (isUserInput){
        if (nowString.length()==0) return;              //避免越界
        if (nowString.charAt(nowString.length()-1)=='.')
            isDotInputed=false;                         //如果删除了小数点，则修改对应的状态
        nowString=nowString.substring(0,nowString.length()-1);
        if (nowString.length()==0) nowString="0";       //避免空串
        refreshAndSync();
    }
    else {                                              //如果是运算结果就全部清零
        allClear(null);
    }
}
```

1.3.4 数值计算方法实现

1. 加减乘除运算

根据记录的数值和运算符对结果进行计算，主要代码如下：

```
void calc(){
    double result=0;
    switch (savedOp){
        case '+':
            result=prevNumber+nowNumber;
            break;
        case '-':
            result=prevNumber-nowNumber;
            break;
        case '*':
            result=prevNumber*nowNumber;
            break;
        case '/':
```

```
            result=prevNumber/nowNumber;
            break;
    }
    TextView history=((TextView) findViewById(R.id.textViewCalcHistory));
    history.setText(
            history.getText()+"\n"+
            Double.toString(prevNumber)+savedOp+Double.toString(nowNumber)+
            "="+Double.toString(result)
    );
    final ScrollView sv1 = (ScrollView) findViewById(R.id.scrollView1);
    sv1.post(new Runnable() {
            public void run() {
                sv1.fullScroll(ScrollView.FOCUS_DOWN);
            }
    });
    nowNumber=result;
}
```

2．开方运算响应方法

对输入的数值进行开方运算，代码如下：

```
public void calcSqrt(View view) {
    isUserInput=false;
    nowNumber=Math.sqrt(nowNumber);
    refreshAndSync();
}
```

3．负数操作

根据原有数值是否有负号进行字符串操作，代码如下：

```
public void toggleSign(View view) {
        if (isUserInput){
            if (nowNumber==0) return;        //避免-0
            if (nowString.charAt(0)=='-')
                nowString=nowString.substring(1, nowString.length());
            else
                nowString='-'+nowString;
        }
        else {
            nowNumber*=-1;
        }
        refreshAndSync();
}
```

1.4 关键知识点解析

1.4.1 在程序中创建菜单

1．创建选项菜单

Android 的 Activity 已经为我们提前创建好了 android.view.Menu 对象，并提供了回调方法 onCreateOptionsMenu(Menu menu)供我们初始化菜单的内容。该方法只会在选项菜单第一次显示的时候被执行，如果需要动态改变选项菜单的内容，请使用 onPrepareOptionsMenu(Menu)。代码如下：

```java
@Override
public boolean onCreateOptionsMenu(Menu menu) {
    //调用父类方法来加入系统菜单
    super.onCreateOptionsMenu(menu);
    //添加菜单项（多种方式）
    //1.直接指定标题
    menu.add("菜单项1");
    //2.通过资源指定标题
    menu.add(R.string.menuitem2);
    //3.显示指定菜单项的组号、ID号、排序号和标题
    menu.add(
        1,              //组号
        Menu.FIRST,     //唯一的ID号
        Menu.FIRST,     //排序号
        "菜单项3");     //标题
    //如果希望显示菜单，请返回true
    return true;
}
```

上面的代码对 Android 添加菜单项的三种方法进行的列举，下面进一步说明第三种方法 add(int groupId, int itemId, int order, CharSequence title)。其中，第一个参数是组号，Android 支持给菜单进行分组，以便方便快速地操作同一组的菜单；第二个参数指定每个菜单项的唯一 ID 号，可以由开发者指定，也可以让系统来自动分配，在响应菜单时用户需要通过 ID 号来判断哪个菜单被点击了，因此常规的做法是定义一些 ID 常量，但在 Android 中有更好的方法，就是通过资源文件来引用，示例工程中就是按这种方式实现的；第三个参数是代表菜单项显示顺序的编号，编号小的显示在前面。

2. 给菜单项分组

```java
@Override
public boolean onCreateOptionsMenu(Menu menu) {
    super.onCreateOptionsMenu(menu);
    //添加4个菜单项，分成两组
    int group1 =1;
    int gourp2 =2;
    menu.add(group1, 1, 1, "item 1");
    menu.add(group1, 2, 2, "item 2");
    menu.add(gourp2, 3, 3, "item 3");
    menu.add(gourp2, 4, 4, "item 4");
    //显示菜单
    return true;
}
```

可以像上面这样给菜单项分组，分组之后就能使用 menu 中提供的方法对组进行操作了，代码如下：

```
menu.removeGroup(group1);                    //删除一组菜单
menu.setGroupVisible(gourp2, visible);       //设置一组菜单是否可见
menu.setGroupEnabled(gourp2, enabled);       //设置一组菜单是否可点
menu.setGroupCheckable(gourp2, checkable, exclusive);   //设置一组菜单的勾选情况
```

3. 响应菜单项

Android 提供了多种响应菜单项的方式，下面逐一介绍。

（1）通过 onOptionsItemSelected()方法响应菜单项。Activity 类中的 onOptionsItemSelected(MenuItem)回调方法是用户经常需要重写的方法之一，每当有菜单项被点击时，Android 就会调用该方法，并传入被点击的菜单项。代码如下：

```
@Override
public boolean onOptionsItemSelected(MenuItem item) {
    switch (item.getItemId()) {
        //响应每个菜单项（通过菜单项的ID号）
        case1:
            //do something here
            break;
        case2:
            //do something here
            break;
        case3:
            //do something here
            break;
        case4:
            //do something here
            break;
        default:    //对于没有处理的事件，交给父类来处理
            return super.onOptionsItemSelected(item);
    }
    //返回true表示处理完菜单项的事件，不需要将该事件继续传播下去了
    return true;
}
```

以上代码可作为用 onOptionsItemSelected()方法响应菜单项的模板来使用，这里为了方便理解，将每个条件（case）后的菜单 ID 使用硬编码形式在程序中展现，但更好的建议是用户可以使用常量或资源 ID 号来使代码更加健壮。

（2）使用监听器响应菜单项。虽然第一种方法是推荐使用的方法，但 Android 还是提供了类似 Java 界面编程的监听器方式来响应菜单项。使用监听器的方式分为两步：创建监听器类和为菜单项注册监听器。

```
//第一步：创建监听器类
    class MyMenuItemClickListener implements OnMenuItemClickListener {
        @Override
        public boolean onMenuItemClick(MenuItem item) {
            //do something here
            return true;     //完成处理
        }
    }
```

//第二步：为菜单项注册监听器
menuitem.setOnMenuItemClickListener(new MyMenuItemClickListener());

Android 文档对 onMenuItemClick(MenuItem item)回调方法的说明是 "Called when a menu item has been invoked. This is the first code that is executed; if it returns true, no other callbacks will be executed."。其含义是当某一个菜单项被调用时，onMenuItemClick()方法就会第一个被执行，如果其返回为 true，则其他回调方法都不会被执行，由此可说明该方法先于 onOptionsItem-Selected()方法执行。

（3）使用 Intent 响应菜单项。第三种方法是直接在 MenuItem 上调用 setIntent(Intent intent)方法，这样 Android 会自动在该菜单被点击时调用 startActivity(Intent)。相对来说，还是直接在 onOptionsItemSelected()的 case 里调用 startActivity(Intent)看起来更为直观。

1.4.2 基础界面布局

Android 的布局负责管理组件的分布和大小，而不是直接设置组件的位置和大小。Android 中所涉及的布局概念主要包括布局文件和布局类型。布局文件是基于 XML 格式编写的，而布局类型主要包括线性布局（LinearLayout）、相对布局（RelativeLayout）、帧布局（FrameLayout）、表格布局（TableLayout）、网格布局（GridLayout）和绝对布局（AbsoluteLayout）。

在所有布局中，比较常用的是线性布局和相对布局。线性布局供给开发者的是一个可以控制的水平或者垂直排列的模型，可以通过重力设置子控件的对齐方式。线性布局不会换行，所有子组件一个挨着一个从头到尾进行排列，排列到头之后，剩下的组件将不会被显示。除了方向和重力，线性布局还有一个非常重要的特性，即权重，将在 1.4.6 节中进行详细讲解。线性布局的重要属性见表 1-1。

表 1-1　线性布局的重要属性

属性名称	对应方法	说明
android:orientation	setOrientation(int)	设置线性布局的排列方式，可取水平（horizontal）和垂直（vertical）
android:gravity	setGravity(int)	设置布局内部组件的对齐方式，支持 top、bottom、left、right、center_vertical、fill_vertical、center_horizontal、fill_horizontal、center、fill、clip_vertical、clip_horizontal 等。也可以同时指定对齐方式的组合，如 right|center_vertical 代表出现在屏幕右侧而且垂直居中，竖线"|"前后不可以出现空格

相对布局中的所有子组件都可以按照相对位置关系进行排列。一个组件位置可以是相对于兄弟组件的位置（如位于兄弟组件的上、下、左、右）或者父容器的位置（如位于父容器的顶部、中央等）。在设计时要按照组件之间的依赖关系排列，由于相对布局的灵活性，在两个控件具有互相指向的规则时，要注意在使用时避免循环依赖的产生。

例如，A 在 B 的左边，B 在 A 的右边，这样两个组件的位置没有指定确定的位置。如果在布局设计中使用了循环依赖，工程将会产生以下错误信息：Illegal State Exception:Circular dependencies cannot exist in a RelativeLayout。

相对布局的父子关系见表 1-2，兄弟关系见表 1-3。

表 1-2 相对布局的父子关系

属性	说明
android:layout_centerHorizontal	子组件在布局中水平居中
android:layout_centerVertical	子组件在布局中垂直居中
android:layout_centerInParent	子组件位于布局中心
android:layout_alignParentBottom	子组件底端与布局底端对齐
android:layout_alignParentLeft	子组件左边与布局左边对齐
android:layout_alignParentRignt	子组件右边与布局右边对齐
android:layout_alignParentTop	子组件顶部与布局顶部对齐

注：属性值必须为 true 或 false

表 1-3 相对布局的兄弟关系

属性	说明
android:layout_toRightOf	组件 A 位于给出 ID 组件的右侧
android:layout_toLeftOf	组件 A 位于给出 ID 组件的左侧
android:layout_above	组件 A 位于给出 ID 组件的上方
android:layout_below	组件 A 位于给出 ID 组件的下方
android:layout_alignTop	组件 A 的上边与给出 ID 组件的上边对齐
android:layout_alignBottom	组件 A 的下边与给出 ID 组件的下边对齐
android:layout_alignLeft	组件 A 的左边与给出 ID 组件的左边对齐
android:layout_alignRight	组件 A 的右边与给出 ID 组件的右边对齐

注：属性值必须为兄弟组件的 ID 号

在 Android 中使用布局可以通过两种方式实现。一种是在工程 res 目录下的 layout 目录中创建布局 XML 文件，之后通过 setContentView(R.layout.<布局文件名称>)方法载入 XML 布局，继而使用 findViewById(R.id.<控件名称>)方法在程序中加载相应的布局或控件。例如，在工程的 res/layout 目录下创建 activity_test_layout.xml 布局文件，内容如下：

```
<LinearLayout xmlns:android="http://schemas.android.com/apk/res/android"
    xmlns:tools="http://schemas.android.com/tools"
    android:layout_width="match_parent"
    android:layout_height="match_parent"
    android:orientation="vertical"
    tools:context=".TestActivity" >
<TextView
        android:id="@+id/textview"
        android:layout_width="wrap_content"
        android:layout_height="wrap_content"
        android:text="@string/hello_world" />
</LinearLayout>
```

继而在代码 TestLayoutActivity.java 中载入布局文件，如下：

```java
public class TestLayoutActivity extends Activity {

    @Override
    protected void onCreate(Bundle savedInstanceState) {
        super.onCreate(savedInstanceState);
        setContentView(R.layout.activity_test_layout);
        TextView tv = (TextView) findViewById(R.id.textview);
    }
}
```

另一种是直接使用代码 TestActivity.java 生成布局的实例,例如:

```java
public class TestActivity extends Activity {
    @Override
    protected void onCreate(Bundle savedInstanceState) {
        super.onCreate(savedInstanceState);
        LinearLayout ll = new LinearLayout(this);
        ll.setOrientation(LinearLayout.VERTICAL);
        ll.setLayoutParams(new LinearLayout.LayoutParams(
                LinearLayout.LayoutParams.MATCH_PARENT,LinearLayout.LayoutParams.MATCH_PARENT));
        TextView tv = new TextView(this);
        tv.setText(R.string.hello_world);
        ll.addView(tv,new LinearLayout.LayoutParams(
                LinearLayout.LayoutParams.WRAP_CONTENT,LinearLayout.LayoutParams.WRAP_CONTENT));
        setContentView(ll);
    }
}
```

1.4.3 设置程序名称和图标

在工程的 AndroidManifest.xml 文件中添加如下语句即可设置程序名称和图标。

```xml
<application
    android:icon="@drawable/ic_launcher"
    android:label="@string/app_name"
>
```

这里需要预先在/res/drawable/目录下放一个名为 ic_launcher.png 的图标图片,并且在/res/values/strings.xml 中定义 app_name 这个字符串(即程序名)。

然后在工程上右击,选择 Refresh(或按 F5 键)刷新工程即可。

1.4.4 常用文本输入控件及按钮

Android 为开发者提供了大量常用控件,如文本框、输入框、按钮、图片框、列表、网格等,我们首先应该掌握文本框(TextView)、输入框(EditText)和按钮(Button)的使用方法。

TextView 用来显示固定长度的文本字符串或者标签,EditText 允许用户编辑文本框中的内容,Button 可供用户点击,触发一个 OnClick 事件,并可设置按钮显示的文本。由于 EditText 和 Button 都继承自 TextView,因此这三个控件拥有很多共用属性,见表 1-4。

表 1-4 TextView、EditText 和 Button 的常用属性

XML 属性	相关方法	说明
android:gravity	setGravity(int)	文本的对齐方式
android:width	setWidth(int)	文本框宽度
android:height	setHeight(int)	文本框高度
android:hint	setHint(int resid)	内容为空时默认显示的提示文本
android:text	setText(int resid)	设置文本显示内容（资源方式）
android:text	setText(CharSequence)	设置文本显示内容（字符串方式）
android:textColor	setTextColor(ColorStateList)	文本颜色
android:textSize	setTextSize(float)	文本大小
android:textStyle	setTextStyle(Typeface)	文本字体风格，如粗体、斜体等
android:scrollHorizontally	setHorizontallyScrolling(boolean)	不够显示全部内容时是否允许水平滚动
android:onClick	setOnClickListener(OnClickListener)	在 XML 中使用 OnClick 属性时，可以为其指定一个方法名，需要在 Java 代码中实现该方法。而使用 setOnClickListener 需要为其提供一个实现了 OnClickListener 接口的类的实例

在 Android 中，所有控件的祖先 View 本身就可以进行 OnClick 事件的监听处理，因此可以为 View 的所有子孙设置 OnClickListener（点击事件监听器）。在实际编码过程中，我们一般为 Button、TextView、ImageView、LinearLayout、RelativeLayout 等组件和布局设置 OnClickListener，其中设置的主要方法一般有以下 4 种：

- 匿名内部类的方式。
- 自定义内部类的方式。
- 由 Activity 实现 View.OnClickListener 接口的方式。
- 在 XML 文件中声明 android:onClick 的方式。

接下来对上述 4 种方式逐一进行介绍，并对比一下在哪种场景下使用哪种方式更好。

1. 匿名内部类的方式

由于为 View 设置 OnClick 事件的实质是为其设置一个实现了 View.OnClickListener 接口的类，因此通过匿名内部类的方式可以使代码更加简洁和紧凑，在需要点击事件较少的 Activity 中使用这种方式更利于代码的连贯性。但要注意的是，如果匿名内部类中需要引用其外部的局部变量时，需要为变量增加 final 修饰符，具体布局和代码如下：

```
<LinearLayout xmlns:android="http://schemas.android.com/apk/res/android"
    xmlns:tools="http://schemas.android.com/tools"
    android:layout_width="match_parent"
    android:layout_height="match_parent"
    android:orientation="vertical"
    android:padding="10dp"
    tools:context="com.example.book.MainActivity" >

    <Button
```

```
        android:id="@+id/button1"
        android:layout_width="wrap_content"
        android:layout_height="wrap_content"
        android:text="Button1" />

    <Button
        android:id="@+id/button2"
        android:layout_width="wrap_content"
        android:layout_height="wrap_content"
        android:text="Button2" />
</LinearLayout>

public class MainActivity extends Activity {
    Button mBtn1,mBtn2;

    @Override
    protected void onCreate(Bundle savedInstanceState) {
        super.onCreate(savedInstanceState);
        setContentView(R.layout.activity_main);
        mBtn1 = (Button) findViewById(R.id.button1);
        mBtn2 = (Button) findViewById(R.id.button2);
        mBtn1.setOnClickListener(new View.OnClickListener() {
            @Override
            public void onClick(View v) {
                //TODO Auto-generated method stub
                Toast.makeText(MainActivity.this, "mBtn1被点击！", Toast.LENGTH_SHORT).show();
            }
        });
        final String message = "mBtn2被点击！";
        mBtn2.setOnClickListener(new View.OnClickListener() {

            @Override
            public void onClick(View v) {
                //TODO Auto-generated method stub
                Toast.makeText(MainActivity.this, message, Toast.LENGTH_SHORT).show();
            }
        });
    }
}
```

2. 自定义内部类的方式

该方式需要在 Activity 中定义一个实现了 View.OnClickListener 接口的内部类，具体代码如下：

```
public class MainActivity extends Activity {
        Button mBtn1,mBtn2;
        class MyClickListener implements View.OnClickListener {
```

```java
        @Override
        public void onClick(View v) {
            //TODO Auto-generated method stub
            switch (v.getId()) {
            case R.id.button1:
                Toast.makeText(MainActivity.this, "mBtn1被点击！", Toast.LENGTH_SHORT).show();
                break;
            case R.id.button2:
                Toast.makeText(MainActivity.this, "mBtn2被点击！", Toast.LENGTH_SHORT).show();
                break;
            default:
                break;
            }
        }
    }
    @Override
    protected void onCreate(Bundle savedInstanceState) {
        super.onCreate(savedInstanceState);
        setContentView(R.layout.activity_main);
        mBtn1 = (Button) findViewById(R.id.button1);
        mBtn2 = (Button) findViewById(R.id.button2);
        mBtn1.setOnClickListener(new MyClickListener());
        mBtn2.setOnClickListener(new MyClickListener());
    }
}
```

3. 由 Activity 实现 View.OnClickListener 接口的方式

该方法通过 Activity 直接实现 View.OnClickListener 接口，同时在 Activity 中实现 onClick(View v) 方法。如果当前布局中具有点击事件的控件很多，利用这种方式将使代码更加简洁直观，但编写代码时需要逐一对应布局 XML 文件中控件的 android:id 属性值。具体代码如下：

```java
public class MainActivity extends Activity implements View.OnClickListener{
    Button mBtn1,mBtn2;

    @Override
    protected void onCreate(Bundle savedInstanceState) {
        super.onCreate(savedInstanceState);
        setContentView(R.layout.activity_main);
        mBtn1 = (Button) findViewById(R.id.button1);
        mBtn2 = (Button) findViewById(R.id.button2);
        mBtn1.setOnClickListener(this);
        mBtn2.setOnClickListener(this);

    }
    @Override
    public void onClick(View v) {
```

```
            //TODO Auto-generated method stub
                switch (v.getId()) {
                case R.id.button1:
                        Toast.makeText(MainActivity.this, "mBtn1被点击！", Toast.LENGTH_SHORT).show();
                    break;
                case R.id.button2:
                        Toast.makeText(MainActivity.this, "mBtn2被点击！", Toast.LENGTH_SHORT).show();
                    break;
                default:
                    break;
                }
            }
        }
```

4. 在 XML 文件中声明 android:onClick 的方式

该方法通过在布局 XML 文件中给控件定义 android:onClick 的属性值来使 Android 通过 Java 的反射机制找到属性值在 Activity 中所对应的同名方法。这种方式很容易理解和学习，即使是没有掌握匿名内部类等 Java 语法的开发者也可以直接使用，但这种方式不是特别直观，后期维护时不利于代码的查找和阅读。

```xml
<LinearLayout xmlns:android="http://schemas.android.com/apk/res/android"
    xmlns:tools="http://schemas.android.com/tools"
    android:layout_width="match_parent"
    android:layout_height="match_parent"
    android:orientation="vertical"
    android:padding="10dp"
    tools:context="com.example.book.MainActivity" >

    <Button
        android:id="@+id/button1"
        android:layout_width="wrap_content"
        android:layout_height="wrap_content"
        android:text="Button1"
        android:onClick="clickBtn1" />

    <Button
        android:id="@+id/button2"
        android:layout_width="wrap_content"
        android:layout_height="wrap_content"
        android:text="Button2"
        android:onClick="clickBtn2" />

</LinearLayout>

public class MainActivity extends Activity implements View.OnClickListener{

        @Override
```

```java
    protected void onCreate(Bundle savedInstanceState) {
        super.onCreate(savedInstanceState);
        setContentView(R.layout.activity_main);
    }

    public void clickBtn1(View v) {
        Toast.makeText(MainActivity.this, "mBtn1被点击！", Toast.LENGTH_SHORT).show();
    }

    public void clickBtn2(View v) {
        Toast.makeText(MainActivity.this, "mBtn2被点击！", Toast.LENGTH_SHORT).show();
    }
}
```

1.4.5 为按钮增加多种样式——selector

很多时候，我们需要按钮在被点击时改变背景或文字颜色，以提供给用户合适的交互反馈。是否被点击，其实是控件不同的状态。在 Android 中可以通过 selector 文件为控件预设在某种状态下的 drawable 或 color，我们可以方便地将其用在根据状态展示不同背景和文字颜色的场景中。在下面的例子中，我们要为 Button 控件分别设置背景 selector 文件和文字颜色 selector 文件。

以下是 btn_bg_selector.xml 的代码，用于设置按钮在不同状态下的背景，当 selector 作为 drawable 资源使用时，应该放置于工程的 res/drawable 目录下，item 必须指定 android:drawable 属性，代码如下：

```xml
<?xml version="1.0" encoding="utf-8"?>
<selector xmlns:android="http://schemas.android.com/apk/res/android">
    <!-- 当前窗口失去焦点时 -->
    <item android:drawable="@drawable/bg_btn_lost_window_focused" android:state_window_focused="false" />
    <!-- 不可用时 -->
    <item android:drawable="@drawable/bg_btn_disable" android:state_enabled="false" />
    <!-- 按压时 -->
    <item android:drawable="@drawable/bg_btn_pressed" android:state_pressed="true" />
    <!-- 被选中时 -->
    <item android:drawable="@drawable/bg_btn_selected" android:state_selected="true" />
    <!-- 被激活时 -->
    <item android:drawable="@drawable/bg_btn_activated" android:state_activated="true" />
    <!-- 默认时 -->
    <item android:drawable="@drawable/bg_btn_normal" />
</selector>
```

以下是 btn_txt_selector.xml 的代码，用于设置按钮在不同状态下的文本颜色，当 selector 作为 color 资源使用时，应该放置于工程的 res/color 目录下，item 必须指定 android:color 属性，代码如下：

```xml
<?xml version="1.0" encoding="utf-8"?>
<selector xmlns:android="http://schemas.android.com/apk/res/android">
    <!-- 当前窗口失去焦点时 -->
```

```xml
    <item android:color="@android:color/black" android:state_window_focused="false" />
    <!-- 不可用时 -->
    <item android:color="@android:color/background_light" android:state_enabled="false" />
    <!-- 按压时 -->
    <item android:color="@android:color/holo_blue_light" android:state_pressed="true" />
    <!-- 被选中时 -->
    <item android:color="@android:color/holo_green_dark" android:state_selected="true" />
    <!-- 被激活时 -->
    <item android:color="@android:color/holo_green_light" android:state_activated="true" />
    <!-- 默认时 -->
<item android:color="@android:color/white" />
</selector>
```

保存好上述 selector 文件后，便可以在 Button 控件中分别直接引用这两个文件，代码如下：

```xml
<Button
    android:id="@+id/btn_login"
    android:layout_width="match_parent"
    android:layout_height="wrap_content"
    android:background="@drawable/btn_bg_selector"
android:text="我是一个按钮"
android:textColor="@color/btn_txt_selector" />
```

接下来对在 selector 文件中支持的状态进行逐一说明。

- android:state_enabled：设置触摸或点击事件是否可用状态，一般只在 false 时设置该属性，表示不可用状态。
- android:state_pressed：设置是否按压状态，一般在 true 时设置该属性，表示已按压状态，默认为 false。
- android:state_selected：设置是否选中状态，true 表示已选中，false 表示未选中。
- android:state_checked：设置是否勾选状态，主要用于 CheckBox 和 RadioButton，true 表示已被勾选，false 表示未被勾选。
- android:state_checkable：设置勾选是否可用状态，类似于 state_enabled，只是 state_enabled 会影响触摸或点击事件，而 state_checkable 则影响勾选事件。
- android:state_focused：设置是否获得焦点状态，true 表示获得焦点，默认为 false，表示未获得焦点。
- android:state_window_focused：设置当前窗口是否获得焦点状态，true 表示获得焦点，false 表示未获得焦点，例如拉下通知栏或弹出对话框时，当前界面就会失去焦点；另外，ListView 的 ListItem 获得焦点时也会触发 true 状态，可以理解为当前窗口就是 ListItem 本身。
- android:state_activated：设置是否被激活状态，true 表示被激活，false 表示未激活，需要 API Level≥11 进行支持，可通过代码调用控件的 setActivated(boolean)方法设置是否激活该控件。
- android:state_hovered：设置是否鼠标在上面滑动的状态，true 表示鼠标在上面滑动，默认为 false，需要 API Level≥14 进行支持。
- android:enterFadeDuration：在状态改变时出现淡入效果。当状态改变时，新状态展示时的淡入时间以毫秒为单位，需要 API Level≥11 进行支持。
- android:exitFadeDuration：在状态改变时出现淡出效果。当状态改变时，旧状态消失时的淡出时间以毫秒为单位，需要 API Level≥11 进行支持。

在 selector 文件的使用过程中，有几点需要格外注意：

（1）selector 作为 drawable 资源时，item 指定 android:drawable 属性，并放于 res/drawable 目录下。

（2）selector 作为 color 资源时，item 指定 android:color 属性，并放于 res/color 目录下。

（3）color 资源也可以放于 drawable 目录下，用@drawable 来引用，但不推荐这么做，drawable 资源和 color 资源最好还是分开。

（4）android:drawable 属性除了引用@drawable 资源外，也可以引用@color 颜色值，但 android:color 只能引用@color。

（5）item 是从上往下匹配的，如果匹配到一个 item，那它就将采用这个 item，而不是采用最佳匹配的规则。所以设置默认的状态一定要写在最后，如果写在前面，则后面所有的 item 都不会起作用。

1.4.6 多分辨率适配利器——LinearLayout

Android 操作系统的大获成功源于其开放性，但开放也带来了大量碎片化的问题，众多的分辨率带给使用者更多的选择，但对开发者提出了更加严格的要求。

Android 为了解决多分辨率适配的问题，提供了多种手段。如在 dimens.xml 中使用 dp 为不同分辨率配置宽高，在 res 资源文件中可以针对不同分辨率设置不同的 layout 和 drawable，使用 Fragment 等。然而在实际开发中，如果针对每种分辨率的设备都提供一种布局和图片，那么不但开发更加繁琐、测试和维护更加困难，就连输出的 apk 安装包的容量也会成倍增加，这样无论是给开发者还是给使用者都带来不小的麻烦。那么有没有一种方式能够更好地解决多分辨率适配的问题呢？下面就来一起研究 Android 为我们提供的 LinearLayout（线性布局）。

作为 Android 中最重要的也是使用频率最高的布局之一，LinearLayout 提供给开发者的是一个可以控制的水平或者垂直排列的模型，方向、重力和权重是 LinerLayout 的首要特性。

1. LinearLayout 的方向（Orientation）

在 LinearLayout 中，布局方向可以分为水平排列和垂直排列。水平排列时占一行，垂直排列时占一列，当水平或者垂直排列超过一行或者一列时，超出屏幕的组件将不会被显示。解决此问题可以使用 ScrollView 或 HorizontalScrollview 组件。

2. LinearLayout 的重力（Gravity）

Android 视图组件中的重力实际上是对齐的概念。由于存在两种十分相似的属性和用法 android:gravity 和 android:layout_gravity，初学者容易混淆，因此需要特别注意。android:gravity 用于设置组件内部的对齐方式。如果该组件是一个容器，比如 LinearLayout，则对齐方式作用于其子组件，即控制子组件在容器中以何种方式对齐；如果该组件是一个非容器的视图控件，比如 TextView，则对齐方式作用于该组件内部的文字或者图片。android:layout_gravity 用于设置组件自身在其父容器中的对齐方式。

有这样一种情况：一个容器控件（如 LinearLayout）设置了 android:gravity 属性对其内部

子组件（如 TextView）的对齐进行约束，而其子组件（TextView）使用 android:layout_gravity 属性安排自身在父容器（LinearLayout）中的对齐方式。通过分析可以得知这样的情况下可能造成对齐冲突，Android 解决冲突的方案是赋予子组件优先权，当子组件通过 android:layout_gravity 安排自己在父容器中的对齐方式时，父容器的 android:gravity 属性将不对该子组件生效。

3. LinearLayout 的权重（Weight）

LinearLayout 的权重是一个非常重要的概念，通过合理地使用权重可以大大增强 Android 应用程序对多种分辨率设备的适配能力。根据经验得知，HTML 网页可以很好地适配不同的设备以进行显示，其中按百分比分配组件宽高的方式功不可没。Android 中并没有显式地使用百分比的形式分配组件宽高的方法，然而通过使用 LinearLayout 的权重，完全可以实现与按百分比分配相同的布局方式。

LinearLayout 的权重可以作为一个参数参与到子组件宽高的计算中，首先需要确定 LinearLayout 的排列方向，当方向为水平时，权重参与每个子组件宽度的计算；当方向为垂直时，权重参与每个子组件高度的计算。假设 LinearLayout 中有 A、B、C 三个子组件，布局中为 A、B、C 指定的权重分别为 Wa、Wb、Wc，宽度分别为 Xa、Xb、Xc，LinearLayout 排列方向为水平，其计算公式可以描述如下：

A 的真实宽度 = Xa+[LinearLayout 的宽度-(Xa+Xb+Xc)]*[Wa/(Wa+Wb+Wc)]

通过上述公式可以推出以下结论：

（1）当 Xa、Xb、Xc 都为 0 时，A 的真实宽度恰好为其权重 Wa 在 LinearLayout 宽度中的占比 Wa/(Wa+Wb+Wc)，权重 Wa 越大，A 的真实宽度越大。

（2）当 Xa、Xb、Xc 都为 match_parent 时，权重 Wa 越大，A 的真实宽度越小。

（3）当 Xa、Xb、Xc 的和小于 LinearLayout 的宽度时，说明 LinearLayout 还有余下的尺寸未被分配，这样通过指定权重，Xa、Xb、Xc 都可以在保证获得布局中指定宽度的同时按权重比例分配到 LinearLayout 余下的尺寸，权重 Wa 越大，分配到的尺寸越多。

因此，一旦开发者掌握了 LinearLayout 权重和方向的使用方法，就可以方便快捷地针对不同分辨率按类似百分比占比的形式进行布局，这样可以实现大多数屏幕分辨率的适配，当然，后续还可以结合不同分辨率下的 drawable 和 dimens.xml 进行微调，以达到满意效果。另外要说明的是，这种方式并不适合手机和平板电脑使用同一套布局，对于手机和平板电脑共用一个 App 的情况，还是建议使用 Android 提供的 Fragment 响应式解决方案。

下面用一个布局的例子来对 LinearLayout 的权重进行说明。

```
<LinearLayout xmlns:android="http://schemas.android.com/apk/res/android"
    xmlns:tools="http://schemas.android.com/tools"
    android:layout_width="match_parent"
    android:layout_height="match_parent"
    android:background="@drawable/menuback"
    android:orientation="vertical"
    tools:context=".MainActivity" >

    <LinearLayout
        android:layout_width="match_parent"
```

```
            android:layout_height="0dp"
            android:layout_weight="4"
            android:background="#aaff0000" />

        <LinearLayout
            android:layout_width="match_parent"
            android:layout_height="0dp"
            android:layout_weight="6"
            android:background="#aa00ff00" />

    </LinearLayout>
```

由于该 LinearLayout 水平排列，当 android:layout_height="0dp"时，恰好权重按百分比分配，android:layout_weight="4"占整个 LinearLayout 布局中 40%的高度，android:layout_weight="6"占整个 LinearLayout 布局中 60%的高度，具体效果如图 1-12 所示。无论屏幕的宽度如何变化，其高度 40%∶60%的比值不变，则控件高度所占比例不变。

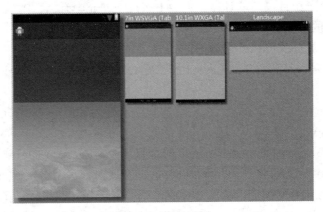

图 1-12　布局示例

1.5　问题与讨论

1．程序如何适应不同的分辨率？

2．按钮有几种不同的状态？如何通过配置 XML 格式布局文件实现在不同状态下显示不同的背景图片？

项目 2　基于离线数据的天气应用——天气预报（一）

在移动应用开发中，天气预报信息是常见的功能。本项目通过对天气预报应用开发过程的学习，进一步掌握移动应用开发的重要思想，在注重项目总体分析、功能模块拆分、操作流程分析、功能及界面设计的同时，进一步增加了程序架构设计的介绍。架构设计在整个应用的实现过程中起到举足轻重的作用，好比建筑大厦的地基，大厦是否稳固、未来是否有改造和升级的空间，都和地基是否合理、健壮息息相关。在学习本项目时，希望读者能够初步体会到架构设计的重要性。

项目需求描述如下：
（1）根据所选城市显示其天气信息。
（2）可以选择并收藏多个城市，选择多个城市后，天气信息可以通过手势在不同城市之间切换。
（3）根据雨、雪、阴、晴等天气情况分别显示不同的图标。

（1）简单程序的架构设计。
（2）掌握 ListView 和 GridView 控件的用法。
（3）掌握 ViewPager 控件的用法。
（4）匿名内部类可以直接访问外部类的成员及其方法的原因。
（5）自定义 Adapter。

2.1　总体设计

2.1.1　总体分析

根据项目需求进行分析，天气预报应用应实现以下功能：界面友好，方便使用，能够显示城市名、当日温度区间、天气情况、实时温度及实时天气的图标，还能够显示未来几天的天气情况；可添加城市，添加新城市后能显示新城市的天气，支持不同城市间天气信息页的切换。

本应用基于离线的天气信息数据，该数据保存在程序代码中。

整个程序除总体模块外，主要分为基础架构模块、用户界面模块和数据管理与控制模块三大部分。在整个系统中，总体模块控制系统的生命周期；用户界面模块负责显示城市的天气数据、天气状态图标以及各个城市间的显示切换；数据管理与控制模块主要提供数据管理功能，为用户界面模块提供数据，同时可以接受并保存用户界面模块产生的数据。

2.1.2 功能模块框图

根据总体分析结果可以总结出功能模块框图，如图 2-1 所示。

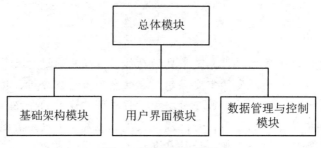

图 2-1　功能模块框图

总体模块的作用主要是生成应用程序的主类，控制应用程序的生命周期；基础架构模块主要提供程序架构、所有 Activity 公用的父类、所有 Activity 公用的方法，包括自定义风格对话框、自定义提示框等功能；数据管理与控制模块主要提供数据获取、数据解析、数据组织和数据缓存功能；用户界面模块包括城市天气信息显示、城市管理显示、操作提示等功能。

2.1.3 系统流程图

根据总体分析结果及功能模块框图梳理出系统启动的主要流程，如图 2-2 所示。

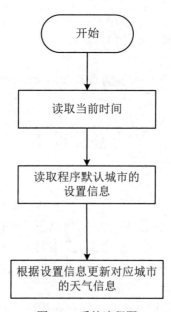

图 2-2　系统流程图

2.1.4 界面设计

在系统总体分析及功能模块划分清楚后,即可开始考虑界面的设计。本应用是显示城市天气情况的应用,设计界面时应该考虑怎样将相关的元素更清晰地表现。

根据程序功能需求可以规划出软件的主要界面,如下:

(1) 启动应用程序界面:启动应用程序的欢迎界面。

(2) 设置界面:对要显示天气预报的城市进行设置。

(3) 显示界面:通过文字和图片显示当前的天气情况,主要包括日期、时间、城市、最高温度、最低温度、当前温度、雨雪情况等,支持不同城市界面的切换,同时每个城市可以显示今后四天的天气情况。

对于设置页,由于主要是城市的选择,因此考虑使用 GridView 方式显示城市列表;对于显示界面,考虑每个城市使用一个屏幕页面,不同屏幕页面间通过滑动切换,因此考虑使用 PagerAdapter 方式显示,一个城市页面中,今后四天的天气情况可以使用 ListView 列表进行展示。

程序界面如图 2-3 所示。

图 2-3 主界面图

从图 2-3 中可以很直观地看到,一个城市天气信息的显示界面区域主要是:当天显示区和未来几天显示区。

- 当天显示区:用于显示当前城市、当前天气情况和当前温度。
- 未来几天显示区:用于显示未来几天的天气情况和温度范围。

可通过滑动屏幕进行不同城市显示页面间的切换。

2.2 详细设计

2.2.1 模块描述

在系统总体分析及界面布局设计完成后,主要工作就转入对各个功能模块的详细设计阶段。

1. 基础架构模块详细设计

基础架构模块主要提供程序架构、所有 Activity 公用的父类、所有 Activity 公用的方法,包括自定义风格对话框、自定义提示框等功能。

基础架构模块功能如图 2-4 所示。

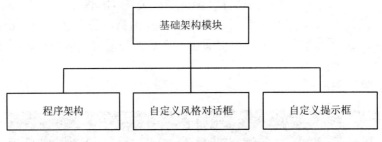

图 2-4 基础架构模块功能图

2. 用户界面模块详细设计

用户界面模块的主要任务是显示天气信息以及实现与用户的交互,即当用户点击按键或者屏幕的时候,监听器会去调用相应的处理方法或其他相应的处理模块。

本模块包括城市天气信息显示、城市管理显示、操作提示等功能。

用户界面模块功能如图 2-5 所示。

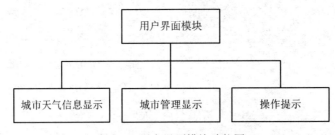

图 2-5 用户界面模块功能图

用户界面模块序列如图 2-6 所示。

3. 数据管理与控制模块详细设计

数据管理与控制模块主要提供数据获取、数据解析、数据组织和数据缓存功能。

数据管理与控制模块和用户界面模块可以调用基础架构模块的一些通用方法。数据管理与控制模块为用户界面模块提供数据,同时可以接受并保存用户界面模块产生的数据。

数据管理与控制模块功能如图 2-7 所示。

项目 2　基于离线数据的天气应用——天气预报（一）

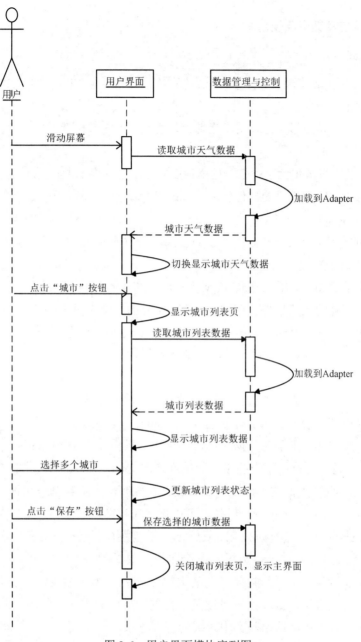

图 2-6　用户界面模块序列图

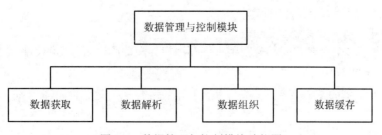

图 2-7　数据管理与控制模块功能图

2.2.2 系统包及其资源规划

1. 文件结构

在系统各个模块的实现方式和流程设计完成后，就可以对系统主要的包和资源进行规划，划分的原则主要是保持各个包相互独立，耦合度尽量低。

在 Android 中，界面部分也采用了当前比较流行的 MVC 框架。

视图（View）层：

在工程的 res 的 layout 包中存放一些 XML 文件以进行界面的描述，使用的时候可以非常方便地引入，将 XML 中描述的控件转换成 Java 对象进行代码级操作。当然，在 Android 中也可以使用 JavaScript 与 HTML 相结合的方式作为 View 层，这样实现需要进行 Java 和 JavaScript 之间的通信，Android 提供了它们之间非常方便的通信实现。

控制（Controller）层：

在 Android 中 Activity 用来连通 View 层的 UI 和 Model 层的数据，通常 Activity 接收、分发、处理用户交互的相关事件，读取并加载 UI 界面，同时对 UI 控件进行控制，向 Model 层发起数据请求，如网络操作、数据库操作等，并接收 Model 层返回的数据，通过 UI 空间反馈给用户，因此 Activity 可以说是控制层的核心组件。

模型（Model）层：

Android 的模型层一般用来进行业务处理，通常网络加载数据、读取文件和数据库、进行复杂运算等事项都在模型层中进行处理，在 com.longdong.studio.section9_weather_b.model 包中存放了模型层文件，用于对数据库和对网络等的操作。在 com.longdong.studio.section9_weather_b.adapter 包中，存放了一些用于数据绑定的文件，在 Android SDK 中的数据绑定也都采用了与 MVC 框架类似的方法来显示数据。在控制层上将数据按照视图模型的要求（也就是 Android SDK 中的 Adapter）封装就可以直接在视图模型上显示了，从而实现了数据绑定。比如显示 Cursor 中所有数据的 ListActivity，其视图层就是一个 ListView，将数据封装为 ListAdapter，并传递给 ListView，数据就在 ListView 中显示。

根据系统功能设计，本系统封装一个基础的 Activity 类，加载各个子页的通用控件，并提供一些基础的实现方法，例如设置进度条、标题等常用的方法，程序中的 Activity 都可继承此基类，继承后就可直接使用基类中封装的基础方法。

系统使用两个 Activity，一个用于显示城市天气信息，一个用于显示城市设置列表。包及其资源结构如图 2-8 所示。

2. 命名空间

本示例设置了多个命名空间，分别用来保存用户界面、后台服务的源代码文件，见表 2-1。

3. 源代码文件

源代码文件及说明见表 2-2。

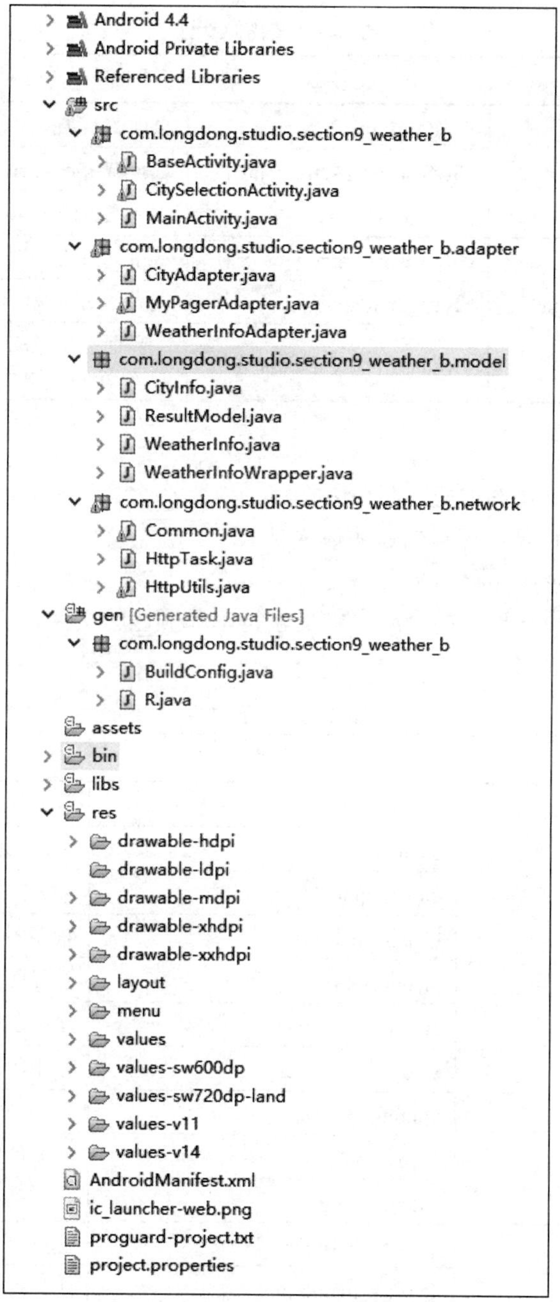

```
> Android 4.4
> Android Private Libraries
> Referenced Libraries
∨ src
  ∨ com.longdong.studio.section9_weather_b
    > BaseActivity.java
    > CitySelectionActivity.java
    > MainActivity.java
  ∨ com.longdong.studio.section9_weather_b.adapter
    > CityAdapter.java
    > MyPagerAdapter.java
    > WeatherInfoAdapter.java
  ∨ com.longdong.studio.section9_weather_b.model
    > CityInfo.java
    > ResultModel.java
    > WeatherInfo.java
    > WeatherInfoWrapper.java
  ∨ com.longdong.studio.section9_weather_b.network
    > Common.java
    > HttpTask.java
    > HttpUtils.java
∨ gen [Generated Java Files]
  ∨ com.longdong.studio.section9_weather_b
    > BuildConfig.java
    > R.java
  assets
> bin
> libs
∨ res
  > drawable-hdpi
    drawable-ldpi
  > drawable-mdpi
  > drawable-xhdpi
  > drawable-xxhdpi
  > layout
  > menu
  > values
  > values-sw600dp
  > values-sw720dp-land
  > values-v11
  > values-v14
  AndroidManifest.xml
  ic_launcher-web.png
  proguard-project.txt
  project.properties
```

图 2-8 包及其资源结构

表 2-1 命名空间

命名空间	说明
com.longdong.studio.section9_weather_b	存放与用户界面相关的源代码文件
com.longdong.studio.section9_weather_b.adapter	存放与页面适配器相关的源代码文件

表 2-2　源代码文件

包名称	文件名	说明
com.longdong.studio.section9_weather_b	BaseActivity.java	页面的基类
	CitySelectionActivity.java	选择城市页的 Activity
	MainActivity.java	主界面页的 Activity
com.longdong.studio.section9_weather_b.adapter	CityAdapter.java	城市列表的 Adapter
	MyPagerAdapter.java	主界面切换城市用 ViewPager 的 Adapter
	WeatherInfoAdapter.java	主界面显示未来几天天气情况的 ListView 的 Adapter，用户界面模块

4. 资源文件

Android 的资源文件保存在/res 的子目录中。

- /res/drawable 目录：保存的是图像文件。
- /res/layout 目录：保存的是布局文件。
- /res/values 目录：保存的是用来定义字符串和颜色的文件。

资源文件及说明见表 2-3。

表 2-3　资源文件

资源目录	文件	说明
drawable	baoyu.png	暴雨的图标
	dayu.png	大雨的图标
	duoyun.png	多云的图标
	ic_launcher.png	程序图标文件
	menuback.jpg	主界面背景图
	qing.png	晴的图标
	xiaoyu.png	小雨的图标
	yin.png	阴的图标
	zhenyu.png	阵雨的图标
	zhongyu.png	中雨的图标
layout	activity_base.xml	基础页的布局
	activity_main.xml	主界面页面的布局
	city_item.xml	城市列表页面的布局
	weather_info_item.xml	列表行的布局
values	arrays.xml	数组声明
	colors.xml	颜色声明
	dimens.xml	屏幕适配相关参数声明
	strings.xml	字符串声明
	styles.xml	样式声明

2.2.3 主要方法流程设计

查看天气流程图如图 2-9 所示。

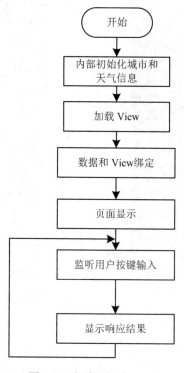

图 2-9 查看天气流程图

2.3 代码实现

2.3.1 显示界面布局

1. 系统主界面

系统主界面是进入系统后显示的界面，该界面包括一个 PageView、一个 ListView、若干 TextView，如图 2-10 所示。

2. 城市设置界面

城市设置界面用于设置在主界面中显示的城市，在此界面中显示了 34 个大城市的名称，用户可以设置选择，该界面包括一个 GridView，如图 2-11 所示。

2.3.2 控件设计实现

1. activity_base.xml

在工程的 res/layout 目录下，创建一个名称为 activity_base.xml 的布局文件，该布局文件采用 RelativeLayout 布局管理。该布局文件中放置了一个 TextView 和两个 Button。TextView

用于显示页面标题，一个 Button 用于显示标题栏左侧的"返回"按钮，默认为隐藏，另一个 Button 用于显示标题栏右侧的"城市"按钮，以显示城市设置页面。

图 2-10　系统主界面　　　　　图 2-11　城市设置界面

具体代码如下：

```
<RelativeLayout xmlns:android="http://schemas.android.com/apk/res/android"
    xmlns:tools="http://schemas.android.com/tools"
    android:layout_width="match_parent"
    android:layout_height="match_parent"
    android:background="#ffffff" >

    <RelativeLayout
        android:id="@+id/top"
        android:layout_width="match_parent"
        android:layout_height="47dp"
        android:background="#666666" >

        <Button
            android:id="@+id/back_btn"
            android:layout_width="wrap_content"
            android:layout_height="wrap_content"
            android:layout_margin="3dp"
            android:textColor="#ffffff"
            android:textSize="14sp"
            android:text="返回"
            android:visibility="gone" />
```

```xml
        <TextView
            android:id="@+id/title_txt"
            android:layout_width="wrap_content"
            android:layout_height="wrap_content"
            android:layout_centerInParent="true"
            android:text="天气预报"
            android:textColor="#ffffff"
            android:textSize="22sp"
            android:visibility="visible" />
        <Button
            android:id="@+id/right_btn"
            android:layout_width="wrap_content"
            android:layout_height="wrap_content"
            android:layout_alignParentRight="true"
            android:layout_margin="3dp"
            android:textColor="#ffffff"
            android:text="城市"
            android:visibility="visible" />
    </RelativeLayout>

    <RelativeLayout
        android:id="@+id/container"
        android:layout_width="match_parent"
        android:layout_height="match_parent"
        android:layout_alignParentBottom="true"
        android:layout_below="@id/top"
        android:background="#cfcfcf" />

</RelativeLayout>
```

2. activity_main.xml

在工程的 res/layout 目录下，创建一个名称为 activity_main.xml 的布局文件，其外层是 LinearLayout，内层嵌套 LinearLayout 以完成布局管理。该布局文件中放置了三个 TextView，分别用于显示城市、天气和实时温度，一个 ListView 用于显示未来几天的天气情况。具体代码如下：

```xml
<LinearLayout xmlns:android="http://schemas.android.com/apk/res/android"
    xmlns:tools="http://schemas.android.com/tools"
    android:layout_width="match_parent"
    android:layout_height="match_parent"
    android:background="@drawable/menuback"
    android:orientation="vertical"
    tools:context=".MainActivity" >

    <LinearLayout
        android:layout_width="match_parent"
        android:layout_height="0dp"
```

```xml
            android:layout_weight="4"
            android:orientation="vertical"
            android:gravity="center">

            <TextView
                android:id="@+id/city"
                android:layout_width="wrap_content"
                android:layout_height="wrap_content"
                android:text="城市"
                android:textColor="#ffffff"
                android:textSize="34sp" />

            <TextView
                android:id="@+id/weather"
                android:layout_width="wrap_content"
                android:layout_height="wrap_content"
                android:text="天气"
                android:textColor="#ffffff"
                android:textSize="22sp" />

            <TextView
                android:id="@+id/cur_temp"
                android:layout_width="wrap_content"
                android:layout_height="wrap_content"
                android:text="实时温度"
                android:textColor="#ffffff"
                android:textSize="52sp" />
        </LinearLayout>

        <ListView
            android:id="@+id/listview"
            android:layout_width="match_parent"
            android:layout_height="0dp"
            android:layout_weight="6"/>
    </LinearLayout>
```

（1）city_item.xml。在工程的 res/layout 目录下，创建一个名称为 city_item.xml 的布局文件。该布局文件里面有一个 TextView，用于显示城市列表页面中的一个城市名称。具体代码如下：

```xml
    <TextView
        xmlns:android="http://schemas.android.com/apk/res/android"
        xmlns:tools="http://schemas.android.com/tools"
        android:id="@+id/city"
        android:layout_width="wrap_content"
        android:layout_height="30dp"
        android:text="城市"
        android:gravity="center"
```

```
        android:padding="10dp"
        android:textColor="#000000"
        android:textSize="18sp" />
```

（2）weather_info_item.xml。在工程的 res/layout 目录下，创建一个名称为 weather_info_item.xml 的布局文件。该布局文件采用 LinearLayout 布局属性，将 orientation 属性设置成 horizontal。该布局文件中放置了两个 TextView 和一个 ImageView。两个 TextView 分别用于显示星期和温度，ImageView 用于显示天气情况图标。具体代码如下：

```
<LinearLayout xmlns:android="http://schemas.android.com/apk/res/android"
    xmlns:tools="http://schemas.android.com/tools"
    android:layout_width="match_parent"
    android:layout_height="match_parent"
    android:orientation="horizontal"
     android:padding="10dp"/>

    <TextView
            android:layout_weight="2"
            android:id="@+id/week"
            android:layout_width="0dp"
            android:layout_height="wrap_content"
            android:text="星期日"
            android:textColor="#ffffff"
            android:textSize="18sp" />
    <ImageView
            android:layout_weight="2"
            android:id="@+id/icon"
            android:layout_width="0dp"
            android:layout_height="wrap_content"
            android:src="@drawable/qing"/>

    <TextView
            android:layout_weight="3"
            android:id="@+id/temperature"
            android:layout_width="0dp"
            android:layout_height="wrap_content"
            android:text="温度"
            android:textColor="#ffffff"
            android:textSize="18sp" />

</LinearLayout>
```

2.3.3 主要代码功能分析

1. BaseActivity 的创建

BaseActivity 作为项目所有 Activity 的基类，需要封装如下内容。

（1）通用的 UI 总布局。由于一个项目中多数页面的风格具有一致性，比如页面的标题栏部分一般放置一条背景、一个标题、左侧一个返回键、右侧可能有一个按钮，如图 2-12 所示，因此需要在基类中对这些可以复用的布局进行统一管理，这样也易于项目的维护。

图 2-12 BaseActivity 中的标题栏部分

（2）Activity 栈的管理。对于早期的 Android 系统（4.0 版之前），开发者经常在 Application 中定义一个用于管理 Activity 的集合，每当一个继承自 BaseActivity 基类的 Activity 执行生命周期中的 onCreate()方法时，都会将这个新创建的 Activity 加入这个集合。这种做法的好处是，可以在全局管理 Activity，也可以方便地找到任意一个 Activity，还可以关闭所有 Activity。当然，自 Android 4.0 以后，系统提供了 registerActivityLifecycleCallbacks()回调方法，Activity 的管理方法可以更方便地实现。

（3）返回键的处理。具有返回键是 Android 操作系统的一大特色。在一般应用中，从首页进入二级页面后，在二级页面左上角会有一个返回键，这个键可以在基类中进行统一处理，如果是首页，则不显示；如果是非首页，则可以将 Activity 从管理集合中弹出（pop），并调用 finish()方法关闭该弹出的 Activity。

（4）通用的进度条和对话框。在一个应用中经常会使用加载进度条或者对话框，这些控件的风格一般都是一致的，因此也非常适合封装在基类中。

代码示例如下：

```
protected void onCreate(Bundle savedInstanceState) {
    super.onCreate(savedInstanceState);
```

```java
// 设置该Activity不显示标题
requestWindowFeature(android.view.Window.FEATURE_NO_TITLE);
// 设置该Activity只能以垂直方向显示
setRequestedOrientation(ActivityInfo.SCREEN_ORIENTATION_PORTRAIT);
getWindow().setSoftInputMode(WindowManager.LayoutParams.SOFT_INPUT_STATE_ALWAYS_HIDDEN);
setContentView(R.layout.activity_base);
// 获取LAYOUT_INFLATER_SERVICE服务，用来加载R.layout.activity_base以外的布局文件
inflater = (LayoutInflater) getSystemService(LAYOUT_INFLATER_SERVICE);
container = (ViewGroup) findViewById(R.id.container);
header = (ViewGroup) findViewById(R.id.top);
title_txt = (TextView) findViewById(R.id.title_txt);
back_btn= (Button) findViewById(R.id.back_btn);
right_btn= (Button) findViewById(R.id.right_btn);
back_btn.setOnClickListener(new View.OnClickListener() {
    @Override
    public void onClick(View v) {
        finish();
        overridePendingTransition(android.R.anim.slide_in_left,
                android.R.anim.slide_out_right);
    }
});
//调用自己的抽象方法，将总布局container提供给子类
onConentViewLoad(container);
}
```

2. BaseActivity 的通用方法封装

```java
//设置页面标题
protected void setTitle(String title) {
    title_txt.setText(title);
}
//显示进度
public void showProgress(String title) {
    if (null == progressDialog) {
        progressDialog = new ProgressDialog(this);
        progressDialog.setProgressStyle(ProgressDialog.STYLE_SPINNER);
        progressDialog.setTitle(title);
        progressDialog.setMessage("请稍候……");
        progressDialog.setCancelable(true);
    }
    dismissProgress();
    progressDialog.show();
}
//关闭进度
public void dismissProgress() {
    if (null != progressDialog) {
        progressDialog.dismiss();
    }
```

```java
        }
        //显示信息提示类型的对话框
        public void showAlertDialog(String title, String message) {
            if(isFinishing()){return;}
            AlertDialog.Builder ab = new AlertDialog.Builder(this);
            ab.setTitle(title)
                    .setMessage(message)
                    .setCancelable(false)
                    .setPositiveButton("确   定",
                            new DialogInterface.OnClickListener() {
                                public void onClick(
                                        DialogInterface dialogInterface, int i) {
                                }
                            });
            ab.create().show();
        }
        //显示可以自定义确定事件的对话框
        public void showAlertDialog(String title, String message,
                DialogInterface.OnClickListener event, String confirm) {
            AlertDialog.Builder ab = new AlertDialog.Builder(this);
            ab.setTitle(title).setMessage(message).setCancelable(false)
                    .setPositiveButton(confirm, event);
            ab.create().show();
        }
        //显示可以自定义确定及取消事件的对话框
        public void showAlertDialogWithCancel(String title, String message,
                DialogInterface.OnClickListener event, String confirm) {
            AlertDialog.Builder ab = new AlertDialog.Builder(this);
            ab.setTitle(title)
                    .setMessage(message)
                    .setCancelable(false)
                    .setPositiveButton(confirm, event)
                    .setNegativeButton("取  消",
                            new DialogInterface.OnClickListener() {
                                @Override
                                public void onClick(DialogInterface dialog,
                                        int which) {
                                    dialog.dismiss();
                                }
                            });
            ab.create().show();
        }
```

3. CitySelectionActivity 的构建

CitySelectionActivity 类继承自 BaseActivity，作用是为用户提供城市选择功能。它支持多个城市的选择，多选功能利用 GridView 控件实现，通过自定义 Adapter 来实现对 CityInfo 中数据的展示。

```java
//直接调用基类的创建函数进行页面的创建
protected void onCreate(Bundle savedInstanceState) {
    super.onCreate(savedInstanceState);
}
//Activity每次初始化显示时进行页面的控件显示设置
protected void onConentViewLoad(ViewGroup container) {
    //TODO Auto-generated method stub
    back_btn.setVisibility(View.VISIBLE);
    right_btn.setVisibility(View.GONE);

    GridView gridView = new GridView(this);
    gridView.setNumColumns(3);
    final CityAdapter adapter = new CityAdapter(this, Common.cityInfoList);
    gridView.setOnItemClickListener(new AdapterView.OnItemClickListener() {
        @Override
        public void onItemClick(AdapterView<?> parent, View view,
                int position, long id) {
            //TODO Auto-generated method stub
            CityInfo info = adapter.getItem(position);
            if(info.isSelected()){
                info.setSelected(false);
                if(checkIsZero()){
                    info.setSelected(true);
                    Toast.makeText(CitySelectionActivity.this, "请至少选择一个城市",
                            Toast.LENGTH_LONG).show();
                }
            }else{
                info.setSelected(true);
            }
            adapter.notifyDataSetChanged();
        }
    });
    //将GridView控件加入BaseActivity的container控件中
    container.addView(gridView);
    gridView.setAdapter(adapter);
}
```

4. MainActivity

```java
//先调用基类的创建函数进行页面的创建，定义右侧按钮的响应方法
protected void onCreate(Bundle savedInstanceState) {
    super.onCreate(savedInstanceState);
    right_btn.setOnClickListener(new View.OnClickListener() {
        @Override
        public void onClick(View v) {
            //TODO Auto-generated method stub
            Intent in = new Intent(MainActivity.this,CitySelectionActivity.class);
            startActivity(in);
```

```java
            }
        });
    }
    //初始化天气数据
    private void initWeatherData(){
        weatherData.clear();
        int j = 0;
        for (CityInfo info:Common.cityInfoList) {
            if(!info.isSelected()){continue;}

            WeatherInfoWrapper wrapper = new WeatherInfoWrapper(info.getCity());
            List<WeatherInfo> list = new ArrayList<WeatherInfo>();
            list.add(new WeatherInfo("今天", j + 1 + " °C", null, null, "晴",
                    "东风微风", "-6°C～-10°C"));
            list.add(new WeatherInfo("周一", j + 2 + " °C", null, null, "阴",
                    "东北风微风", "-7°C～-10°C"));
            list.add(new WeatherInfo("周二", j + 3 + " °C", null, null, "小雨",
                    "南风微风", "-1°C～-9°C"));
            list.add(new WeatherInfo("周三", j + 4 + " °C", null, null, "中雨",
                    "西风微风", "-1°C～-5°C"));
            list.add(new WeatherInfo("周四", j + 5 + " °C", null, null, "多云",
                    "东风微风", "0°C～-5°C"));

            wrapper.setWeatherInfoList(list);
            weatherData.add(wrapper);
        }
    }
```

5. common.java

通常 common 类用于放置一些公用的变量或者静态常量，一般在初始化集合类型的静态常量时经常需要向集合中添加一些初始数据。这里给大家介绍一个小技巧，就是 Java 的双括号语法，具体代码示例如下：

```java
public static final String[] CITIES =  {"天津","上海","重庆","石家庄","郑州","武汉","长沙","南京","南昌",
"沈阳","长春","哈尔滨","太原","西安","广州","济南","成都","西宁","海口","合肥","贵阳","杭州","福州",
"台北","兰州","昆明","拉萨","银川","南宁","乌鲁木齐","呼和浩特","香港","澳门"};
    //使用双括号语法初始化currentCitySet
    //将CityInfo结构初始化并增加到列表cityInfoList中
    public static   List<CityInfo> cityInfoList =new ArrayList<CityInfo>(){
        {     add(new CityInfo(DEFAULT_CITY, true));
            for(String city :CITIES){
                add(new CityInfo(city, false));
            }
        }
    };
```

双括号语法的原理其实并不难懂：第一层括号实际上是定义了一个匿名内部类（Anonymous Inner Class），第二层括号实际上是一个实例初始化块（Instance Initializer Block），这个块在匿名内部类构造时被执行。这个块之所以被叫做"实例初始化块",是因为它们被定义在了一个类的实例范围内。虽然这种方式很方便,但是要注意,由于匿名内部类使用不当可能会引起内存溢出,因此,在开发者对 Java 匿名内部类相关机制还没有深入透彻地理解前,双括号语法这种方式更适合配合 static 修饰符来初始化静态常量,而应尽量避免在复杂的业务逻辑代码中使用这种语法。另外初始化静态常量还可以使用 static 代码块的方式,请对比如下代码中 DATA1 和 DATA2 的初始化过程:

```
public static final Map<String, String> DATA1 = new HashMap<String, String>()
    {{
        this.put("key1", "value1");
        this.put("key2", "value2");
    }};

public static final Map<String, String> DATA2 = new HashMap<String, String>();
static
{
    DATA2.put("key1", "value1");
    DATA2.put("key2", "value2");
}
```

2.4 关键知识点解析

对于 Android 常用的基本界面组件,如 Button、TextView、EditText、ImageView 等,这些组件用于显示基本的数据单位,只要对每个控件的布局属性和方法加以了解,就可以较快速地掌握其用法。然而,当数据是以数组或集合形式表示时,我们将如何处理呢?诚然,可以利用循环的方式将数据逐一显示在各个基本界面组件中,可是当数组或集合中的数据改变时,我们需要界面组件跟踪这些改变,并同步更改显示的数据。显然,仅通过循环加载基本组件的方式来处理数据的改变并不轻松,因此,Android 专门提供了相应的组件来加载数组或集合,同时使用 Adapter 适配器的方案来动态管理数据变更与 View 显示的同步。本节我们将从学习 ListView 开始,对 Android 中 AdapterView 系列组件和 Adapter 进行全面的了解。

2.4.1 ListView 控件的用法

1. ListView 使用的基本步骤

(1) 得到 ListView 类型的对象 mListView。
(2) 生成适配器对象 mListViewAdapter,并给该适配器对象设置数据。
- 使用 SimpleAdapter 类。
- 使用 ArrayAdapter 类。
- 自己写适配器:继承 BaseAdapter 并重写其中的方法。

(3) 调用 ListView 类型对象的.setAdapter 方法把 mListViewAdapter 对象设置为 mListView 的适配器。
(4) 调用 mListView 的监听方法设置各种监听事件。

2. 具体步骤和代码

第一步：得到 ListView 类型的对象 listView。

在布局文件中添加如下代码，也就是在布局文件中加入 ListView 控件，然后再在 Activity 中调用 findViewById(R.id.mListView)方法得到 mListView 对象。

```
<ListView
        android:id="@+id/mListView"
        android:layout_width="fill_parent"
    android:layout_height="fill_parent"/>
```

第二步：生成适配器对象 mListViewAdapter 并设置数据。

可以根据需求，从下面三种类型中选择其一生成适配器对象 mListViewAdapter。

（1）ArrayAdapter 适配器所生成的列表：每一个项里有且只有一个 TextView 控件。

（2）SimpleAdapter 适配器所生成的列表：每一项中可以有任意多个控件，控件类型可以包括 TextView、ImageView、CheckBox 等。

（3）继承 BaseAdapter 并重写各方法所生成的适配器类：更加灵活多变，可按照自己的需求随意控制每一项的布局。

ArrayAdapter 适配器的使用：

（1）把要在列表每一项中显示的文字按顺序放在一个数组中（如下代码中的 arrayString）。

（2）使用 ArrayAdapter 类的构造方法创建 ArrayAdapter 类型的适配器对象。

1）第一个参数为 Context 类型，这里写 this。

2）第二个参数为每一个 item 的布局文件的资源 ID，我们在 layout 文件夹中创建了一个 XML 文件 item_marrayadapter.xml，注意在该资源文件中只能包含一个 TextView 控件。

3）第三个参数是要显示的文字的数组，这里写 arrayString。

```
String[] arrayString = {"item1","item2","item3"};
ArrayAdapter arrayAdapter = new ArrayAdapter(
            this,
            R.layout.item_marrayadapter,
            arrayString);
```

item_marrayadapter.xml

```
<?xml version="1.0" encoding="utf-8"?>
<TextView xmlns:android="http://schemas.android.com/apk/res/android"
    android:layout_width="match_parent"
    android:layout_height="match_parent"
/>
```

效果如图 2-13 所示。

SimpleAdapter 的使用：通过 SimpleAdapter 的构造方法 SimpleAdapter(Context context, List<? extends Map<String, ?>> data,int resource, String[] from, int[] to);来创建 SimpleAdapter 适配器对象。

1）第一个参数为 Context 类型，此处写 this。

2）第二个参数为 List<? extends Map<String,?>>类型的对象，此处放的是要在列表每一项中显示的数据。这个对象可以理解为把一个 Map<String,?>类型的对象放在一个 ArrayList<>

中，每一个 Map<String,?>类型的对象也可以就是某一行中要显示的数据，比如下面的例子，通过 for 循环生成了 10 个 map 对象，也就是说在列表中有 10 项，每一个 map 对应一项的数据。

图 2-13　效果图

3）第三个参数为每一个 item 的布局文件的资源 ID。

4）第四个参数和第五个参数是把上面 map 中的数据和资源文件联系起来的关键。第四个参数为 map 中键的数组，第五个参数为布局文件资源中控件的 ID 的数组，也就是说，以第四个参数的数组中第一个值作为键找到在 map 中对应的值，把这个值给第五个参数的数组中第一个值对应的控件，依此类推，就是把 firstNameTextView 键对应的值给 R.id.firstName-TextView，代码如下：

```
List<Map<String,String>> list = 
    new ArrayList<Map<String,String>>();
for(int i = 0 ; i < 10;i ++ )
{
    Map<String,String> map = new HashMap<String, String>();
    map.put("firstNameTextView", "firstNameTextView"+i);
    map.put("secondNameTextView", "secondNameTextView"+i);
    list.add(map);
}
SimpleAdapter simpleAdapter = new SimpleAdapter(
    this,
    list,
    R.layout.item_msimpleadapter,
    new String []{"firstNameTextView","secondNameTextView"},
    new int[]{R.id.firstNameTextView,R.id.secondNameTextView});
```

效果如图 2-14 所示。

第三步：把 mListViewAdapter 对象设置为 mListView 的适配器。

调用 mListView 对象的.setAdapter 方法设置适配器，将适配器对象和 ListView 控件关联。

mListView.setAdapter(simpleAdapter);

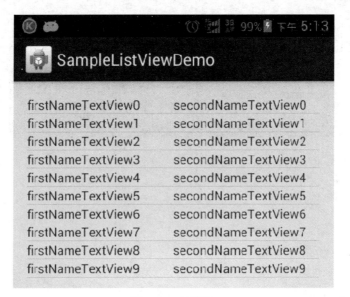

图 2-14　效果图

第四步：设置监听器监听事件的发生。
ListView 对象常用的监听事件的方法如下：
setOnItemClickLinstener()   //某一项被点击
setOnItemLongClickListener()   //某一项被长按
```
    mListView.setOnItemClickListener(new AdapterView.OnItemClickListener() {
        @Override
        public void onItemClick(AdapterView<?> parent, View view,
                int position, long id) {
            System.out.println("第"+position+"项被点击");

        }
    });
mListView.setOnItemLongClickListener(new AdapterView.OnItemLongClickListener() {
        @Override
        public boolean onItemLongClick(AdapterView<?> parent, View view,
                int position, long id) {
            System.out.println("第"+position+"项被长按");
            return false;
        }
    });
```

2.4.2　自定义适配器

在上一节中，我们直接使用了系统提供的 SimpleAdapter 作为 ListView 的适配器。其实 Android 提供的 SimpleAdapter、ArrayAdapter 等 Adapter 的具体实现类在很大程度上可以方便开发，但是如果列表的每一个项目中需要显示的内容都比较复杂，比如同时包含图片、文本、编辑框、多选按钮等，那么利用 Android 提供的具体 Adapter 实现类可能会力不从心，这时候需要我们自定义 Adapter。自定义 Adapter 的过程并不复杂，我们只需要继承 BaseAdapter 类并实现以下 4 种方法：

```
public int getCount()
public Object getItem(int position)
public long getItemId(int position)
public View getView(int position, View convertView, ViewGroup parent)
```

其中，getCount()方法用于返回所适配的数组或集合数据的总数，getItem()方法用于返回数组或集合数据的第 position 项，如果数据中的项目存在 ID，可以使用 getItemId()方法返回数组或集合数据的第 position 项的 ID。在自定义适配器的过程中，需要对 getView()方法进行特别说明，作用是将数组或集合的每一项与 Android 的界面组件相绑定，Android()的 AdapterView 在生成界面的每一项时都会调用一次 getView()方法，getView()方法则返回已装配好数据的界面组件。在 getView()方法中，我们通常使用 LayoutInflater 类的 inflate()方法将布局文件中的组件加载到程序中，继而将数据设置给相应的组件。下面的代码是一个自定义 Adapter 的实例。

XML 布局文件（res/layout/adapter_item.xml）中的代码如下：

```xml
<LinearLayout xmlns:android="http://schemas.android.com/apk/res/android"
    xmlns:tools="http://schemas.android.com/tools"
    android:layout_width="match_parent"
    android:layout_height="match_parent"
    android:orientation="vertical"
    android:padding="15dp" >
    <ImageView
        android:id="@+id/avatar"
        android:layout_width="wrap_content"
        android:layout_height="wrap_content"
        android:layout_marginRight="20dp"
        android:src="@drawable/ic_launcher" />

    <TextView
        android:id="@+id/nickname"
        android:layout_width="wrap_content"
        android:layout_height="wrap_content"
        android:layout_toRightOf="@id/avatar"
        android:text="昵称"
        android:textSize="22sp" />

    <TextView
        android:id="@+id/gender"
        android:layout_width="wrap_content"
        android:layout_height="wrap_content"
        android:layout_marginLeft="20dp"
        android:layout_toRightOf="@id/nickname"
        android:text="性别"
        android:textSize="22sp" />
</LinearLayout>
```

集合中的对象类型为 UserInfoBean，其代码如下：

```java
public class UserInfoBean {

    private String nickname;
    private int gender ;
    private String photo;

    public String getGenderStr() {
        if(gender==0){
            return "女";
        }else{
            return "男";
        }

    }

    public String getNickname() {
        return nickname;
    }

    public void setNickname(String nickname) {
        this.nickname = nickname;
    }

    public String getPhoto() {
        return photo;
    }

    public void setPhoto(String photo) {
        this.photo = photo;
    }

    public int getGender() {
        return gender;
    }

    public void setGender(int gender) {
        this.gender = gender;
    }
}
```

自定义的 Adapter 为 MyAdapter，其代码如下：

```java
public class MyAdapter extends BaseAdapter {
    Context context;
    List<UserInfoBean> list ;
    LayoutInflater inflater;
```

```java
    public MyAdapter(Context context, List<UserInfoBean> list) {
        super();
        this.context = context;
        this.list = list;
        inflater = (LayoutInflater) context.getSystemService(Context.LAYOUT_INFLATER_SERVICE);
    }

    @Override
    public int getCount() {
        return list.size();
    }

    @Override
    public UserInfoBean getItem(int position) {
        return list.get(position);
    }

    @Override
    public long getItemId(int position) {
        return 0;
    }

    @Override
    public View getView(int position, View convertView, ViewGroup parent) {
        if(convertView == null){
            convertView = inflater.inflate(R.layout.adapter_item,null );
        }

        TextView nickname_txt = (TextView) convertView.findViewById(R.id.nickname);
        nickname_txt.setText(list.get(position).getNickname());

        TextView gender_txt = (TextView) convertView.findViewById(R.id.gender);
        gender_txt.setText(list.get(position).getGenderStr());

        ImageView avatar = (ImageView) convertView.findViewById(R.id.avatar);
        Bitmap bmp = BitmapFactory.decodeFile(list.get(position).getPhoto());
        avatar.setImageBitmap(bmp);
        return convertView;
    }
}
```

通过上述方式我们就可以实现自定义的 Adapter。不过细心的读者会发现，如果列表中的数据众多，每次 getView 时都要进行大量的 findViewById 操作，这样需要多次从 XML 中读取数据并转化成内存中的 View，比较耗费时间。因此我们在此基础上进一步对代码进行优化，即使用 ViewHolder 机制对已经通过 findViewById 得到的 View 进一步重用，经改造后的代码如下：

```java
public class MyAdapter extends BaseAdapter {
```

```java
Context context;
List<UserInfoBean> list ;
LayoutInflater inflater;


public MyAdapter(Context context, List<UserInfoBean> list) {
    super();
    this.context = context;
    this.list = list;
    inflater = (LayoutInflater) context.getSystemService(Context.LAYOUT_INFLATER_SERVICE);
}

@Override
public int getCount() {
    return list.size();
}

@Override
public UserInfoBean getItem(int position) {
    return list.get(position);
}

@Override
public long getItemId(int position) {
    return 0;
}

@Override
public View getView(int position, View convertView, ViewGroup parent) {
    ViewHolder holder;
    if(convertView == null){
        convertView = inflater.inflate(R.layout.adapter_item,null );
        holder = new ViewHolder();
        holder.nickname_txt = (TextView) convertView.findViewById(R.id.nickname);
        holder.gender_txt = (TextView) convertView.findViewById(R.id.gender);
        holder.avatar = (ImageView) convertView.findViewById(R.id.avatar);
        //setTag用于缓存holder
        convertView.setTag(holder);
    }else {
        //如果缓存的holder可用，则通过getTag取回
        holder = (ViewHolder) convertView.getTag();
    }

    //直接获取holder中的控件，而不需要再通过findViewById获取
    holder.nickname_txt.setText(list.get(position).getNickname());
    holder.gender_txt.setText(list.get(position).getGenderStr());
```

```
            Bitmap bmp = BitmapFactory.decodeFile(list.get(position).getPhoto());
            holder.avatar.setImageBitmap(bmp);
            return convertView;
    }

    static class ViewHolder {
            TextView nickname_txt;
            TextView gender_txt ;
            ImageView avatar;
    }
}
```

2.4.3 GridView 控件的用法

GridView 与 ListView 的使用方式没有本质上的区别，只是在 ListView 垂直排列的基础上为每行增加了水平排列的项目，即先按照每行预设的项目进行水平排列，排满后再进行换行，重复进行水平排列，依此类推。这里需要了解 GridView 不同于 ListView 的几个属性，见表 2-4。

表 2-4 GridView 属性

属性名称	属性值示例	说明
android:numColumns=""	auto_fit	列数设置为自动，亦可设置为具体数字，代表每行显示的项目个数
android:columnWidth=""	90dp	每列的宽度，也就是 Item 的宽度
android:stretchMode=""	columnWidth	缩放与列宽大小同步
android:verticalSpacing=""	10dp	垂直边距
android:horizontalSpacing=""	10dp	水平边距

掌握了上述几个属性后，就可以按照与 ListView 绑定 Adapter 类似的步骤对 GridView 进行相应设置和操作了。

2.4.4 ViewPager 控件的用法

随着 Android 的不断发展，新的优秀控件层出不穷，为了兼容比较早的 Android 版本，谷歌经常为开发者提供软件兼容包，这样就可以在低版本中使用高版本控件的新特性。ViewPager 就是其中一例，它可以为我们提供多个视图左右切换的效果，不但支持通过按钮进行切换，而且支持通过手势进行切换。

使用 ViewPager 和使用其他列表类控件的步骤基本一致，具体如下：

（1）将适配器控件放置在某个布局中。

```
<LinearLayout xmlns:android="http://schemas.android.com/apk/res/android"
    xmlns:tools="http://schemas.android.com/tools"
    android:layout_width="match_parent"
    android:layout_height="match_parent"
    android:orientation="vertical">
```

```xml
<android.support.v4.view.ViewPager
    android:id="@+id/viewpager"
    android:layout_width="match_parent"
    android:layout_height="wrap_content" />
</LinearLayout>
```

（2）为控件中的每一个 item 创建布局。

1）布局 1。

```xml
<LinearLayout xmlns:android="http://schemas.android.com/apk/res/android"
    xmlns:tools="http://schemas.android.com/tools"
    android:layout_width="match_parent"
    android:layout_height="match_parent"
    android:orientation="vertical">
    <TextView
        android:layout_width="wrap_content"
        android:layout_height="wrap_content"
        android:text="布局1" />
</LinearLayout>
```

2）布局 2。

```xml
<LinearLayout xmlns:android="http://schemas.android.com/apk/res/android"
    xmlns:tools="http://schemas.android.com/tools"
    android:layout_width="match_parent"
    android:layout_height="match_parent"
    android:orientation="vertical">
    <TextView
        android:layout_width="wrap_content"
        android:layout_height="wrap_content"
        android:text="布局1" />
</LinearLayout>
```

3）布局 3。

```xml
<LinearLayout xmlns:android="http://schemas.android.com/apk/res/android"
    xmlns:tools="http://schemas.android.com/tools"
    android:layout_width="match_parent"
    android:layout_height="match_parent"
    android:orientation="vertical">
    <TextView
        android:layout_width="wrap_content"
        android:layout_height="wrap_content"
        android:text="布局1" />
</LinearLayout>
```

（3）准备数据集合，加载页卡。

```java
//通过LayoutInflater加载布局
LayoutInflater inflater = (LayoutInflater)
    getSystemService(Context.LAYOUT_INFLATER_SERVICE);
LinearLayout layout1 = inflater.inflate(R.layout.layout1, null);
LinearLayout layout2 = inflater.inflate(R.layout.layout2, null);
```

```
        LinearLayout layout3 = inflater .inflate(R.layout.layout3, null);
        //将加载好的布局放置到集合中
        List<View> itemList = new ArrayList<View>();
        itemList.add(layout1);
        itemList.add(layout2);
        itemList.add(layout3);
```
（4）构造适配器。
```
        class MyPagerAdapter extends PagerAdapter {
            @Override
            public boolean isViewFromObject(View arg0, Object arg1) {
                    return arg0 == arg1;
        }
            @Override
            public int getCount() {
                    return itemList.size();
        }
            @Override
            public void destroyItem(ViewGroup container, int position, Object object) {
                container.removeView(itemList.get(position));
        }
            @Override
            public int getItemPosition(Object object) {
                 return super.getItemPosition(object);
        }
        @Override
        public Object instantiateItem(ViewGroup container, int position) {
                container.addView(itemList.get(position));
                //可以在container中通过findViewById处理相应控件
                return itemList.get(position);
        }
    };
```
（5）将列表控件与适配器（Adapter）相绑定。
```
viewPager viewPager = findViewById(R.id.viewpager);
viewPager.setAdapter(new MyPagerAdapter());
viewPager.setCurrentItem(0);
```
（6）为列表控件设置监听器。
```
    public class MyOnPageChangeListener implements OnPageChangeListener{
            @Override
            public void onPageScrollStateChanged(int arg0) {
        }
            @Override
            public void onPageScrolled(int arg0, float arg1, int arg2) {
        }
            @Override
            public void onPageSelected(int position) {
                    Toast.makeText(TestLayoutActivity.this,
```

"这是第"+ viewPager.getCurrentItem()+"张页卡", Toast.LENGTH_LONG).show();
            }
        }

2.4.5　ListView 中存在按钮时导致 ListItem 点击无效的解决方案

当 ListView 或 GridView 的 Item 中嵌套 Button 或 CheckBox 之类的控件时，Item 本身的焦点会优先被按钮控件抢占，导致 Item 本身的触摸点击事件无效。要避免此种情况的发生，一般有以下三种解决方案：

方案一：将 Button 或 CheckBox 换成 TextView 或 ImageView 之类的控件。
方案二：设置 Button 或 CheckBox 之类的控件的 focusable 属性为 false。
方案三：设置 Item 的根布局属性为 android:descendantFocusability="blocksDescendants"。

在这里，我们更推荐第三种方案，其设置方法快捷方便。当 android:descendantFocusability 属性为 blocksDescendants 时，意味着该布局下的所有子控件都不能获取焦点，这样便巧妙地解决了按钮控件与 List 中的 Item 焦点冲突的问题。

需要提醒读者的是，android:descendantFocusability 属性需要设置给 ViewGroup 类型的控件，在实践中更多的是设置给常用的布局，如 LinearLayout 和 RelativeLayout，其属性值可以有以下三种：

- beforeDescendants：ViewGroup 将使其子控件优先获取到焦点。
- afterDescendants：仅当 ViewGroup 的子类控件不需要获取焦点时，ViewGroup 才获取焦点。
- blocksDescendants：ViewGroup 会阻止其所有子控件获取焦点，而自身直接获得焦点。

2.5　问题与讨论

1. Android 中的 View 类、Graphics 类和 Canvas 类之间是什么关系？
2. 自定义控件的实现方式有哪几种？如何实现？
3. 为什么匿名内部类可以直接访问外部类的成员及其方法？

项目 3 天天爱读书手机阅读器

在企业级移动应用中，文本文件的浏览和展示是十分常见的功能，配合丰富的辅助功能菜单和手势操作，基本可以达到掌上阅读器的效果。在本项目中，我们要学习完整的手机阅读器应用开发，进一步掌握自定义 UI、手势、文件读写、菜单等功能的实现方法，为 Android 技术进阶打下基础。

项目需求描述如下：

（1）SD 卡文件列表界面，列出 SD 卡中存在的所有资料。

（2）从 SD 卡中加载文本文件，生成阅读界面。

（3）支持用手势进行翻页，能够进行动态运算，根据所在页数即时加载该页需要展示的文本内容。

（4）提供辅助功能菜单，用户可以使用文字大小调节、亮度调节、更换背景、收藏等功能。

（1）学习 Touch 事件的相关处理。

（2）掌握自定义控件。

（3）熟悉 Canvas。

3.1 总体设计

本系统是一个电子书阅读器，其核心要求如下：

（1）主界面用来实现阅读电子书的功能。

（2）在主界面进行阅读时，可通过手势进行上下页翻页。

（3）点击"导入书籍"选项可在存储器中查找电子书并加载至主界面。

（4）点击"设置"选项会有多种操作可供选择，包括朗读、设置背景音乐、设置背景图片、设置亮度、设置字体颜色、设置字体大小、设置书签和跳转，用户可通过自己的喜好来设置不同阅读界面的风格，并且可以选择自己喜爱的音乐作为背景音乐以便在阅读时进行播放。

（5）设置书签选项有三种操作可供选择，分别为添加书签、选择书签和删除书签，其中点击书签列表中的某条书签记录对应的删除图标，可达到删除该条记录的目的。

3.1.1 功能模块框图

功能模块框图如图 3-1 所示。

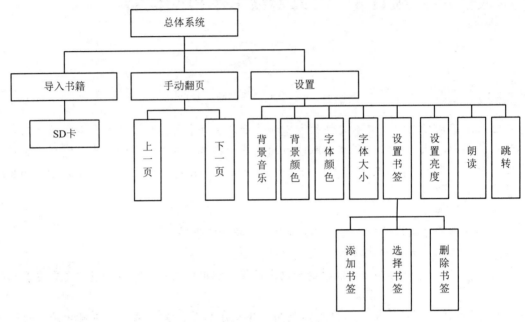

图 3-1　功能模块框图

3.1.2 系统流程图

根据总体分析结果及功能模块框图梳理出系统启动的主要流程，如图 3-2 所示。

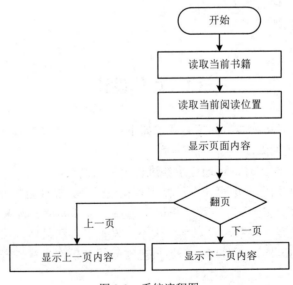

图 3-2　系统流程图

3.1.3 界面设计

根据程序功能需求可以规划出软件的主要界面如下：
- 书籍列表界面：书籍列表显示，可进行阅读设置。
- 阅读界面：显示书籍详细内容页面，支持滑动翻页。
- 设置界面：考虑支持字体颜色、字体大小、背景颜色、背景音乐的设置。

程序主界面如图 3-3 所示。

 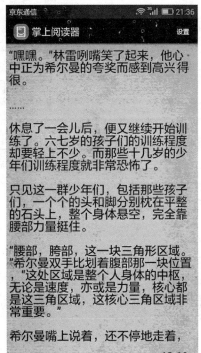

图 3-3　程序主界面图

3.2　详细设计

3.2.1　模块描述

在系统总体分析及界面布局设计完成后，主要工作就转入对各个功能模块的详细设计阶段。

1. 基础架构模块详细设计

基础架构模块主要提供程序架构、所有 Activity 公用的父类、所有 Activity 公用的方法，包括自定义风格对话框、自定义提示框等功能。

基础架构模块功能如图 3-4 所示。

2. 用户界面模块详细设计

用户界面模块的主要任务是显示书籍列表和详细页，以及实现与用户的交互，即当用户点击按键或者屏幕的时候，监听器会去调用相应的处理办法或其他相应的处理模块。

本模块包括书籍列表显示、详细阅读、设置管理等功能。

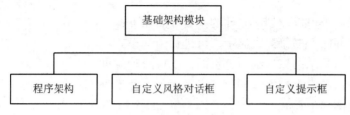

图 3-4　基础架构模块功能图

用户界面模块功能如图 3-5 所示。

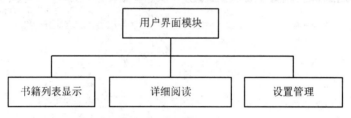

图 3-5　用户界面模块功能图

3.2.2　系统包及其资源规划

1. 文件结构

在系统各个模块的实现方式和流程设计完成后，就可以对系统主要的包和资源进行规划，划分的原则主要是保持各个包相互独立，耦合度尽量低。

系统使用两个 Activity，一个用于显示书籍列表，一个用于显示书籍详细内容页。包及其资源结构如图 3-6 所示。

图 3-6　包及其资源结构

2. 命名空间

本示例设置了多个命名空间，分别用来保存用户界面、后台服务的源代码文件，具体说明见表3-1。

表 3-1 命名空间

命名空间	说明
com.booktest	书籍列表和详细页
com.booktest.adapter	适配器
com.booktest.helper	数据库操作
com.booktest.mydialog	自定义对话框
com.booktest.pref	定义数据
com.booktest.state	状态记录
com.booktest.util	拼音操作
com.booktest.widget	页面操作

3. 源代码文件

源代码文件及说明见表3-2。

表 3-2 源代码文件

包名称	文件名	说明
com.booktest	BookListActivity.java	书籍列表页面
	ImportBookActivity.java	导入书籍页面
	ReadingActivity.java	阅读书籍内容页
com.booktest.adapter	LocAdapter.java	书籍列表适配器
	MarkAdapter.java	书签适配器
com.booktest.helper	LocalBookHelper.java	数据库操作
	MarkHelper.java	数据库操作
com.booktest.mydialog	MarkDialog.java	我的书签的自定义对话框
com.booktest.pref	Consts.java	变量定义
	SettingPrefActivity.java	设置页面
com.booktest.state	BookVo.java	记录书的地址及导入状态
	MarkVo.java	记录书签的各种数据
com.booktest.util	PinyinListComparator.java	List 拼音比较器
	PinyinUtil.java	拼音工具类
com.booktest.widget	BookPageFactory.java	页面控制工厂
	PageWidget.java	控制页面画出贝塞尔曲线

4. 资源文件

Android 的资源文件保存在 /res 的子目录中。

- /res/drawable 目录：保存的是图像文件。
- /res/layout 目录：保存的是布局文件。
- /res/values 目录：保存的是用来定义字符串和颜色的文件。

资源文件及说明见表 3-3。

表 3-3 资源文件

资源目录	文件	说明
drawable	bg.jpg	阅读背景图片
	bg2.jpg	阅读背景图片
	bg3.jpg	阅读背景图片
	bg4.jpg	阅读背景图片
	book0.png	书籍图标
	icon.png	程序图标
	mymarktitle.png	书签列表标题图标
	tool11.png	字号设置背景
	tool22.png	夜间模式设置背景
	tool33_1.png	添加书签按钮
	tool33_2.png	我的书签按钮
layout	bookpop.xml	设置条布局
	in.xml	导入书籍列表
	item_mark.xml	管理标签条目
	main.xml	书籍列表
	mymark.xml	书签列表
	tool11.xml	字号设置弹出页面
	tool22.xml	夜间模式弹出页面
	tool33.xml	添加书签弹出页面
	tool44.xml	跳转设置弹出页面
values	strings.xml	常用字符串定义
	umeng_analyse_strings.xml	常用字符串定义

3.2.3 主要方法流程设计

用户在阅读界面时的操作流程如图 3-7 所示。

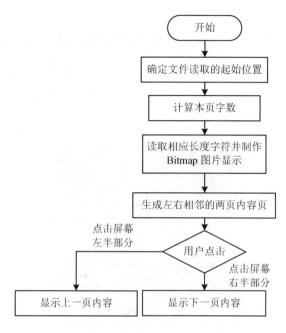

图 3-7　用户在阅读界面时的操作流程图

3.3　代码实现

3.3.1　显示界面布局

1. 书籍选择列表页面

书籍选择列表页面是进入系统后显示的界面，该界面包括两个 ImageView 和一个 ListView。在 ListView 中，列表的每一行包括一个 ImageView 和若干 TextView，如图 3-8 所示。

图 3-8　系统主界面

2. 书籍阅读界面

书籍阅读界面中包括一个 RelativeLayout，程序动态显示书籍内容，如图 3-9 所示。

3.3.2 Touch 事件方法实现

Android 的 Touch 事件处理比较复杂，本节只介绍其具体实现，对于 Touch 事件的详细说明请见 3.4.1 节。

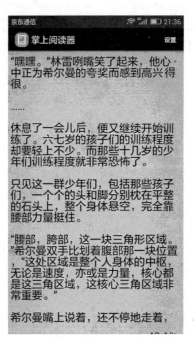

图 3-9　书籍阅读界面

程序根据用户点击页面的位置确定是上翻一页还是下翻一页。

```
mPageWidget = new PageWidget(getApplicationContext(), screenWidth, readHeight);    //页面
mPageWidget.setBitmaps(mCurPageBitmap, mCurPageBitmap);

mPageWidget.setOnTouchListener(new OnTouchListener() {
    @Override
    public boolean onTouch(View v, MotionEvent e) {
        boolean ret = false;
        if (v == mPageWidget) {
            if (!show) {

                if (e.getAction() == MotionEvent.ACTION_DOWN) {
                    if (e.getY() > readHeight) {   //若超出范围，则不翻页
                        return false;
                    }
                    mPageWidget.abortAnimation();
                    mPageWidget.calcCornerXY(e.getX(), e.getY());
                    pagefactory.onDraw(mCurPageCanvas);
```

```
            if (mPageWidget.DragToRight()) {    //左翻
                try {
                    pagefactory.prePage();
                    begin = pagefactory.getM_mbBufBegin();      //获取当前阅读位置
                    word = pagefactory.getFirstLineText();      //获取当前阅读位置的首行文字
                } catch (IOException e1) {
                    Log.e(TAG, "onTouch->prePage error", e1);
                }
                if (pagefactory.isfirstPage()) {
                    Toast.makeText(getApplicationContext(), "当前是第一页",
                        Toast.LENGTH_SHORT).show();
                    return false;
                }
                pagefactory.onDraw(mNextPageCanvas);
            } else {       //右翻
                try {
                    pagefactory.nextPage();
                    begin = pagefactory.getM_mbBufBegin();   //获取当前阅读位置
                    word = pagefactory.getFirstLineText();   //获取当前阅读位置的首行文字
                } catch (IOException e1) {
                    Log.e(TAG, "onTouch->nextPage error", e1);
                }
                if (pagefactory.islastPage()) {
                    Toast.makeText(getApplicationContext(), "已经是最后一页了",
                        Toast.LENGTH_SHORT).show();
                    return false;
                }
                pagefactory.onDraw(mNextPageCanvas);
            }
            mPageWidget.setBitmaps(mCurPageBitmap, mNextPageBitmap);
        }
        editor.putInt(bookPath + "begin", begin).commit();
        ret = mPageWidget.doTouchEvent(e);
        return ret;
    }
  }
  return false;
 }
});
```

3.4 关键知识点解析

3.4.1 Android 的 Touch 事件处理机制

Android 系统原生支持触屏操作，当用户触摸屏幕时，将产生点击事件，即 Touch 事件。每个 Touch 事件会包含一些与触屏相关的信息（如触摸在屏幕上发生的位置、触摸发生的时间等），这些信息被被封装成 MotionEvent 对象，根据用户点击操作动作的不同，Android 系统将 Touch 事件分为 4 种类型，详见表 3-4。

表 3-4 Touch 事件类型

事件类型	描述
MotionEvent.ACTION_DOWN	按下 View
MotionEvent.ACTION_UP	在 View 上抬起
MotionEvent.ACTION_MOVE	在 View 上滑动
MotionEvent.ACTION_CANCEL	非人为原因的结束

在 Android 系统中，Activity 作为控制器是首先接收到 Touch 事件的组件，继而分发给 ViewGroup，ViewGroup 再根据具体情况分发到其子控件（View），通常用户的点击行为都是从点击（ACTION_DOWN）开始，抬起（ACTION_UP）结束，中间可能伴随一次或多次滑动（ACTION_MOVE），这些事件构成了一组连贯的 Touch 动作。对这些连贯动作需要先进行接收、分发，继而进行处理，下面对此进行说明。

Activity 会通过 dispatchTouchEvent()方法将点击事件分发，首先分发给 Activity 中的根 ViewGroup，ViewGroup 再通过其自身的 dispatchTouchEvent()方法将点击事件继续分发给其子 View，View 中的 dispatchTouchEvent()方法将被调用，此时 View 中的 OnTouchEvent(event)方法将被执行，该方法即是 View 具体处理 Touch 事件的方法。通常该方法中需要分别对上表中的 MotionEvent.ACTION_DOWN、MotionEvent.ACTION_UP、MotionEvent.ACTION_MOVE、MotionEvent.ACTION_CANCEL 四种事件进行处理。

3.4.2 掌握自定义控件

在 Android 中自定义控件，一般需要如下步骤：

（1）自定义 View 的属性。需要在 res/values/下建立一个 attrs.xml，在其中定义属性并声明样式。

```
<?xml version="1.0" encoding="utf-8"?>
<resources>
    <attr name="mText" format="string" />
    <attr name="mTextColor" format="color" />
    <attr name="mTextSize" format="dimension" />
    <declare-styleable name="MyCustomView">
        <attr name="mText" />
        <attr name="mTextColor" />
        <attr name="mTextSize" />
    </declare-styleable>
</resources>
```

其中 format 属性为取值的类型，目前 Android 支持 10 种类型：string、color、demension、integer、enum、reference、float、boolean、fraction、flag。

（2）在布局文件中使用上述自定义的属性。
```
<RelativeLayout xmlns:android="http://schemas.android.com/apk/res/android"
    xmlns:tools="http://schemas.android.com/tools"
    xmlns:custom="http://schemas.android.com/apk/res/com.example.view"
    android:layout_width="match_parent"
    android:layout_height="match_parent" >
    <com.example.view.MyCustomView
        android:layout_width="100dp"
        android:layout_height="60dp"
        custom:mText="你好"
        custom:mTextColor="#000000"
        custom:mTextSize="20sp" />
</RelativeLayout>
```
注意这里必须要引入 xmlns:custom="http://schemas.android.com/apk/res/com.example.view" 这个命名空间，后面的包路径指的是项目的包名。

（3）在 View 的构造方法中获得 attr.xml 中自定义的属性。
```
public MyCustomView(Context context, AttributeSet attrs, int defStyle)
{
    super(context, attrs, defStyle);
    TypedArray a = context.getTheme().obtainStyledAttributes(attrs,
    R.styleable.MyCustomView, defStyle, 0);
    int n = a.getIndexCount();
    for (int i = 0; i < n; i++)
    {
        int attr = a.getIndex(i);
        switch (attr)
        {
        case R.styleable.MyCustomView_mText:
            mText = a.getString(attr);
            break;
        case R.styleable.MyCustomView_mTextColor:
            //默认颜色设置为黑色
            mTextColor = a.getColor(attr, Color.BLACK);
            break;
        case R.styleable.MyCustomView_mTextSize:
            //默认设置为16sp，TypeValue也可以把sp转化为px
            mTextSize = a.getDimensionPixelSize(attr, (int) TypedValue.applyDimension(
                    TypedValue.COMPLEX_UNIT_SP, 16, getResources().getDisplayMetrics()));
            break;

        }

    }
    a.recycle();
    mPaint = new Paint();
```

```
        mPaint.setTextSize(mTextSize);
        mPaint.setColor(mTextColor);
        mBound = new Rect();
        mPaint.getTextBounds(mText, 0, mText.length(), mBound);
}
```

（4）重写 onMesure。

```
@Override
    protected void onMeasure(int widthMeasureSpec, int heightMeasureSpec)
    {
        int widthMode = MeasureSpec.getMode(widthMeasureSpec);
        int widthSize = MeasureSpec.getSize(widthMeasureSpec);
        int heightMode = MeasureSpec.getMode(heightMeasureSpec);
        int heightSize = MeasureSpec.getSize(heightMeasureSpec);
        int width;
        int height ;
        if (widthMode == MeasureSpec.EXACTLY)
        {
            width = widthSize;
        } else
        {
            mPaint.setTextSize(mTextSize);
            mPaint.getTextBounds(mTitle, 0, mTitle.length(), mBounds);
            float textWidth = mBounds.width();
            int desired = (int) (getPaddingLeft() + textWidth + getPaddingRight());
            width = desired;
        }

        if (heightMode == MeasureSpec.EXACTLY)
        {
            height = heightSize;
        } else
        {
            mPaint.setTextSize(mTextSize);
            mPaint.getTextBounds(mTitle, 0, mTitle.length(), mBounds);
            float textHeight = mBounds.height();
            int desired = (int) (getPaddingTop() + textHeight + getPaddingBottom());
            height = desired;
        }
        setMeasuredDimension(width, height);
    }
```

（5）重写 onDraw。

```
@Override
    protected void onDraw(Canvas canvas)
    {
        mPaint.setColor(Color.BLUE);
```

```
canvas.drawRect(0, 0, getMeasuredWidth(), getMeasuredHeight(), mPaint);
mPaint.setColor(mTextColor);
canvas.drawText(mText, getWidth() / 2 - mBound.width() / 2,
getHeight() / 2 + mBound.height() / 2, mPaint);
    }
```

3.5 问题与讨论

在 AndroidMenifest.xml 中使用什么标签来设置 ContentProvider？

项目 4　基于本地图库的图片应用——幻彩手机相册

在企业级 Android 应用开发中，针对图片的处理和优化是必不可少的功能，也是初学者经常遇到且无从下手的难点。熟练掌握图片的各种操作及优化方法，可以使应用的体验和性能大幅提升。因此，作为 Android 应用 UI 开发技术中的重中之重，我们在本项目中要学习完整的图库应用开发，进一步掌握图片的加载、剪裁、缩放、压缩等功能的实现方法。

项目需求描述如下：
（1）实现浏览手机相册中的图片。
（2）实现图片切换特效。
（3）实现图片旋转和缩放。
（4）实现媒体播放器。

（1）学习读取图库中图片的方法。
（2）图片处理方法。
（3）媒体播放器的调用。

4.1　总体设计

本系统是一个本地图库的图片应用程序，主要包含图片浏览界面、设置界面、图片编辑界面。图片浏览界面用来实现读取本地相册，进行图片展示。在设置界面中，可以设置幻灯片切换时间、自动隐藏 ActionBar 时间。在图片编辑界面中，可以对图片进行旋转、滤镜、调节、修剪、美肤等操作。

4.1.1　功能模块框图

功能模块框图如图 4-1 所示。

4.1.2　系统流程图

根据总体分析结果及功能模块框图梳理出系统启动的主要流程，如图 4-2 所示。

项目 4　基于本地图库的图片应用——幻彩手机相册

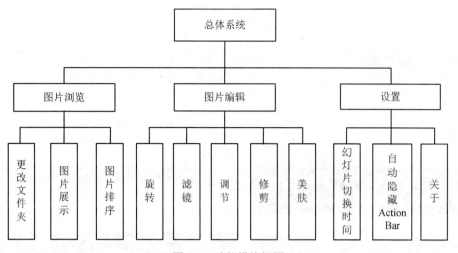

图 4-1　功能模块框图

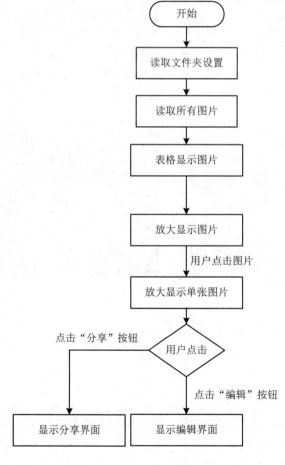

图 4-2　系统流程图

4.1.3 界面设计

根据程序功能需求可以规划出软件的主要界面,如下:
- 图片列表界面:列表显示图片,可以更改图片文件夹、阅读设置。
- 详细查看界面:显示图片详细页面,支持滑动翻页。
- 设置界面:考虑支持幻灯片切换时间、自动隐藏 ActionBar。

程序主界面如图 4-3 所示。

图 4-3　程序主界面

4.2　详细设计

4.2.1 模块描述

在系统总体分析及界面布局设计完成后,主要工作就转入对各个功能模块的详细设计阶段。

1. 基础架构模块详细设计

基础架构模块主要提供程序架构、所有 Activity 公用的父类、所有 Activity 公用的方法,包括自定义风格对话框、自定义提示框等功能。

基础架构模块功能如图 4-4 所示。

项目 4　基于本地图库的图片应用——幻彩手机相册

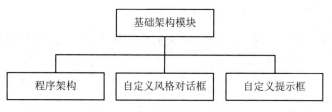

图 4-4　基础架构模块功能图

2. 用户界面模块详细设计

用户界面模块的主要任务是显示图片列表和详细页，以及实现与用户的交互，即当用户点击按键或者屏幕的时候，监听器会去调用相应的处理办法或其他相应的处理模块。

本模块包括图片列表、详细阅览、设置管理等功能。

用户界面模块功能如图 4-5 所示。

图 4-5　用户界面模块功能图

4.2.2　系统包及其资源规划

1. 文件结构

在系统各个模块的实现方式和流程设计完成后，就可以对系统主要的包和资源进行规划，划分的原则主要是保持各个包相互独立，耦合度尽量低。

系统使用两个 Activity，一个用于显示图片列表信息，一个用于显示图片详细信息。包及其资源结构如图 4-6 所示。

图 4-6　包及其资源结构

2. 命名空间

本示例设置了多个命名空间，分别用来保存用户界面、后台服务的源代码文件，具体说明见表 4-1。

表 4-1 命名空间

命名空间	说明
cn.com.cucsi.android.app	该包下放置应用程序主要 Activity
cn.com.cucsi.android.app.directory	该包下放置文件浏览器的相关代码
cn.com.cucsi.android.app.gallery.cursor	该包下放置照片管理器的数据适配器
cn.com.cucsi.android.app.imagedetail	该包下放置图片查看相关代码
cn.com.cucsi.android.app.queries	该包下放置数据查询相关代码
cn.com.cucsi.android.util	该包下放置重新实现的部分 Android 原生类
cn.com.cucsi.android.widget	该包下放置自定义小控件
cn.com.cucsi.database	数据库操作
cn.com.cucsi.io	该包下放置文件输入输出操作相关代码
cn.com.cucsi.util	该包下放置通用工具类

3. 源代码文件

源代码文件及说明见表 4-2。

表 4-2 源代码文件

包名称	文件名	说明
cn.com.cucsi.android.app	AndroFotoFinderApp.java	应用程序的主入口 Application
	BookmarkController.java	书签管理
	Common.java	常用常量声明
	GalleryFilterActivity.java	相册过滤相关代码
	Global.java	应用全局变量声明
	MainActivity.java	应用程序的主 Activity
	OnGalleryInteractionListener.java	相册交互过程的监听器
	SettingsActivity.java	应用设置
cn.com.cucsi.android.app.directory	DirectoryGui.java	文件浏览界面
	DirectoryListAdapter.java	文件列表适配器
	DirectoryLoaderTask.java	文件加载的异步任务
	DirectoryPickerFragment.java	文件选择的 Fragment
cn.com.cucsi.android.app.gallery.cursor	GalleryCursorAdapter.java	相册展示的适配器
	GalleryCursorFragment.java	相册展示的 Fragment
cn.com.cucsi.android.app.imagedetail	ImageDetailActivityViewPager.java	图片显示的 ViewPager 控件实现类
	ImageDetailDialogBuilder.java	图片显示的对话框实现类

续表

包名称	文件名	说明
	ImagePagerAdapterFromCursor.java	图片显示的 ViewPager 控件的适配器
	LockableViewPager.java	图片显示的 ViewPager 控件实现类
cn.com.cucsi.android.app.queries	FotoSql.java	图片查询的 SQL 语句
	FotoViewerParameter.java	用于查询的相关参数
	Queryable.java	用于查询存储类
	SqlJobTaskBase.java	查询的异步任务
cn.com.cucsi.android.util	AndroidFileCommands.java	文件操作相关命令
	ExifGps.java	图片 Exif 信息处理
	GarbageCollector.java	垃圾回收
	GraphUtils.java	常用的图形相关操作工具
	IntentUtil.java	常用的 Intent 相关操作工具
	LogCat.java	重新封装的 Logcat 类
	MediaScanner.java	重新封装的 MediaScanner 类
	RecursiveMediaScanner.java	可递归的 MediaScanner 类
	SelectedFotos.java	已选择的图片信息存储
cn.com.cucsi.android.widget	AboutDialogPreference.java	对话框形式的设置信息
	Dialogs.java	重新封装的对话框类
	DualListAdapter.java	双列表适配器
	EditTextPreferenceWithSummary.java	支持摘要的文本编辑控件
	HistoryEditText.java	记录历史的文本编辑控件
cn.com.cucsi.database	QueryParameter.java	查询相关的存储类
	SelectedItems.java	已选择项的存储类
cn.com.cucsi.io	Directory.java	目录存储类
	DirectoryBuilder.java	目录创建类
	DirectoryFormatter.java	文件格式化工具
	DirectoryNavigator.java	目录导航
	FileCommands.java	文件操作相关命令
	FileUtils.java	文件工具类
	IGalleryFilter.java	图片过滤接口
cn.com.cucsi.util	IsoDateTimeParser.java	时间解析器
	StringTemplateEngine.java	文本模板解析引擎

4. 资源文件

Android 的资源文件保存在/res 的子目录中。
- /res/drawable 目录：保存的是图像文件。
- /res/layout 目录：保存的是布局文件。
- /res/values 目录：保存的是用来定义字符串和颜色的文件。

资源文件及说明见表 4-3。

表 4-3 资源文件

资源目录	文件	说明
	ic_launcher.png	程序主图标
	image_loading.png	加载过场
layout	activity_dir_choose.xml	目录选择布局
	activity_gallery.xml	相册基本布局
	activity_gallery_filter.xml	相册过滤布局
	activity_image_view_pager.xml	图片显示布局
	dialog_about.xml	关于信息的对话框布局
	dialog_edit_name.xml	名称编辑的对话框布局
	dialog_scanner_status.xml	查询器状态对话框布局
	fragment_directory.xml	文件展示的 Fragment 布局
	fragment_gallery.xml	相册展示的 Fragment 布局
	grid_item_gallery.xml	相册展示的网格适配器子项布局
	list_item_directory_child.xml	目录展示列表的二级子项布局
	list_item_directory_parent.xml	目录展示列表的一级子项布局
values	arrays.xml	数组声明
	colors.xml	颜色声明
	dimens.xml	屏幕适配相关参数声明
	strings.xml	字符串声明
	styles.xml	样式声明

4.2.3 主要方法流程设计

用户在阅览界面时的操作流程如图 4-7 所示。

项目 4　基于本地图库的图片应用——幻彩手机相册

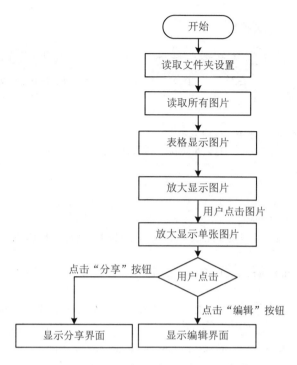

图 4-7　用户在阅览界面时的操作流程图

4.3　代码实现

4.3.1　显示界面布局

系统主界面是进入系统后显示的界面。该界面包括若干 ImageView，如图 4-8 所示。

图 4-8　系统主界面

4.3.2 读取手机图库方法实现

访问系统图库首先需要通知系统负责处理用户选择图库的 Activity，因此需要使用 Intent。这里需要说明的是，需要对版本小于 19 和大于等于 19 的 Intent 分别进行处理。

```java
private void selectPicFromLocal() {
    Intent intent;
    if (Build.VERSION.SDK_INT < 19) {
        intent = new Intent(Intent.ACTION_GET_CONTENT);
        intent.setType("image/*");
    } else {
        intent = new Intent(
                Intent.ACTION_PICK,
                android.provider.MediaStore.Images.Media.EXTERNAL_CONTENT_URI);
    }
    startActivityForResult(intent, PICK_IMAGE_REQUEST_CODE);
}
```

执行完上述代码后，系统将开启用户选择图库的 Activity 供用户选择图片，选择完毕后，需要回调 onActivityResult 方法，因此接下来要学习 onActivityResult 的写法。

```java
protected void onActivityResult(int requestCode, int resultCode, Intent data) {
    super.onActivityResult(requestCode, resultCode, data);
    if (resultCode == RESULT_OK) { //清空消息
        if (requestCode == PICK_IMAGE_REQUEST_CODE) { //发送本地图片

            if (data != null) {
                Uri selectedImage = data.getData();
                String path = null;
                if (selectedImage != null) {
                    Cursor cursor = getContentResolver().query(
                            selectedImage, null, null, null, null);

                    if (cursor != null) {
                        cursor.moveToFirst();
                        int columnIndex = cursor.getColumnIndex("_data");
                        String picturePath = cursor.getString(columnIndex);
                        cursor.close();
                        cursor = null;

                        if (picturePath == null
                                || picturePath.equals("null")) {
                            Toast toast = Toast.makeText(this, "未找到图片",
                                    Toast.LENGTH_SHORT);
                            toast.setGravity(Gravity.CENTER, 0, 0);
                            toast.show();
                            return;
                        }
```

```
                    path = picturePath;
                } else {
                    File file = new File(selectedImage.getPath());
                    if (!file.exists()) {
                        Toast toast = Toast.makeText(this, "未找到图片",
                                Toast.LENGTH_SHORT);
                        toast.setGravity(Gravity.CENTER, 0, 0);
                        toast.show();
                        return;
                    }

                    path = file.getAbsolutePath();
                }
            }
        }
    }
}
```

这里的 path 变量保存的便是用户从系统图库里选择的图片的具体路径，接下来开发者便可以自行对图片进行处理了。

4.3.3 图片方向的判断

某些时候我们会发现系统相机拍照后，读取到的图片在 ImageView 中显示的方向不正确，这时候需要对图片方向进行判断并加以调整。一般手机拍照保存的图片格式为 jpg，Android 可以通过使用 ExifInterface 类对图片方向进行判断，继而进行调整，代码实现如下：

```
Bitmap bm = loadBitmap(imgpath);
int digree = 0;
ExifInterface exif = null;
try {
    exif = new ExifInterface(imgpath);
} catch (IOException e) {
    e.printStackTrace();
    exif = null;
}
if (exif != null) {
    //读取相机中图片的方向信息
    int ori = exif.getAttributeInt(ExifInterface.TAG_ORIENTATION,
            ExifInterface.ORIENTATION_UNDEFINED);
    //计算旋转角度
    switch (ori) {
    case ExifInterface.ORIENTATION_ROTATE_90:
        digree = 90;
        break;
```

```
        case ExifInterface.ORIENTATION_ROTATE_180:
            digree = 180;
            break;
        case ExifInterface.ORIENTATION_ROTATE_270:
            digree = 270;
            break;
        default:
            digree = 0;
            break;
    }
}
if (digree != 0) {
    //旋转图片
    Matrix m = new Matrix();
    m.postRotate(digree);
    bm = Bitmap.createBitmap(bm, 0, 0, bm.getWidth(),bm.getHeight(), m, true);
}
```

4.3.4 图片压缩

当 Android 将图片加载至内存中时,开发者必须要考虑图片在内存中所占的空间,同时为了避免内存溢出,图片使用后应第一时间释放,一旦加载的图片超过内存能容纳的限度,应用就会发生内存溢出而崩溃。在这里介绍一下 Android 中常用的图片压缩方法。

通常,内存中位图(Bitmap)的大小=图片长度×图片宽度×一个像素点占用的字节数,无论是减少图片长度、宽度还是减少每像素占用的字节,都可以缩小位图的尺寸,因此首先尝试对每像素占用的字节数进行处理,这就需要我们了解 Android 中位图的格式。

Android 中有四种常见位图格式,详见表 4-4。

表 4-4 Android 的四种常见位图格式

名称	位深	位图格式	说明
ALPHA_8	8 位	Alpha 位图	该格式下 1 个像素点占用 1 个字节,它没有颜色,只有透明度
ARGB_4444	16 位	ARGB 位图	该格式下 1 个像素点占 4+4+4+4=16 位,2 个字节
RGB_8888	32 位	ARGB 位图	该格式下 1 个像素点占 8+8+8+8=32 位,4 个字节
RGB_565	16 位	RGB 位图	该格式下没有透明度,1 个像素点占 5+6+5=16 位,2 个字节

从表 4-4 中可以分析出,ARGB_8888 格式的位图质量最高,RGB_565 格式的位图次之,因此我们可以考虑将图片压缩为 RGB_565 格式,代码如下:

```
BitmapFactory.Options options = new BitmapFactory.Options();
options.inPreferredConfig = Bitmap.Config.RGB_565;
Bitmap bm = BitmapFactory.decodeFile(
```

```
Environment.getExternalStorageDirectory().getAbsolutePath()
    + "/DCIM/Camera/test.jpg", options);
Log.i("压缩后图片的大小", (bm.getByteCount() / 1024 / 1024)
    + " 宽度:" + bm.getWidth() + " 高度:" + bm.getHeight());
```

存储成 RGB_565 格式的图片比 ARGB_8888 格式的图片大小减少了一半，那是否可以使用另外两种更小的压缩方式呢？答案是不可行的，因为 ARGB_4444 和 ALPHA_8 固然可以占用更小的存储空间，但图片效果已经无法让我们接受了，所以 RGB_565 是一个比较折中的方案，但细心的读者可能已经发现，RGB_565 格式是在忽略了 Alpha 透明通道的情况下降低存储占用的，如果我们的图片必须使用透明通道，即图片中拥有透明或半透明的部分时，使用 RGB_565 这种格式就会丢失透明度信息。

刚刚介绍了通过改变位图格式来对图片进行压缩的方式，接下里我们介绍另一个 Android 中的有损质量压缩的方式，其原理与上述改变压缩格式的方法相近。在该方式压缩过程中，图片的宽、高都没有变，因为质量压缩不会减少图片的像素，它在保持像素的前提下改变图片的位深及透明度等信息来达到压缩图片的目的。其实位深就是存储图片中每个像素所用的位数，它决定了图片每个像素可能存在的最大色彩数，位深越大，所能展现的色彩数越多，或者说色彩越丰富。通过质量压缩后，位深可能被降低，同时透明度信息可能丢失，因此质量压缩可以较大程度降低图片占用的存储空间，但同时也牺牲了图片的质量。在 Android 中，主要通过 Bitmap 实例的 compress 方法实现质量压缩，其中 quality 参数决定图片质量的好坏和压缩率的高低，其参考代码如下：

```
int quality = 95;
Bitmap mBitmap= BitmapFactory.decodeFile(
    Environment.getExternalStorageDirectory().getAbsolutePath());
ByteArrayOutputStream baos = new ByteArrayOutputStream();
mBitmap.compress(CompressFormat.jpeg, quality, baos);
byte[] bytes = baos.toByteArray();
Bitmap mBitmap1= BitmapFactory.decodeByteArray(bytes, 0, bytes.length);
Log.i("压缩后图片的存储大小=", (mBitmap1.getByteCount() / 1024 / 1024)
    + " 宽度:" + mBitmap1.getWidth() + " 高度:" + mBitmap1.getHeight()
    + "二进制Byte数组大小= " + (bytes.length / 1024) + "KB");
```

如果我们持续观察 bytes.length，会发现随着 quality 变小 bytes.length 也在变小。也就是图片在二进制状态尺寸变小了，因此这种压缩方式非常适合存储和传递二进制的图片数据，比如微信分享图片要求传入二进制数据过去，其限制是小于 32KB，而通过这种质量压缩，可以起到减小图片二进制大小的作用。

注意：上述图片压缩的代码实例中 compress 方法的第一个参数使用的是 CompressFormat.jpeg 有损压缩方式，如果我们将 compress 方法中的第一个参数改为 CompressFormat.png 格式，即 mBitmap.compress(CompressFormat.png, quality, baos);，这说明图片将保存为 png 格式，此时 quality 是没有作用的，因为 png 格式采用的是无损压缩方式，bytes.length 不会变化。

4.3.5 使用 Android 提供的媒体播放器（MediaPlayer）

Android 提供的媒体播放器封装了播放音频和视频的能力，可以满足应用中基本的音视频播放需求，将调用过程分解为如下步骤：

1. 创建 MediaPlayer 实例

可以使用直接 new 的方式，代码如下：

MediaPlayer mp = new MediaPlayer();

也可以使用 create 的方式，代码如下：

MediaPlayer mp = MediaPlayer.create(this, R.raw.test_media_file_rsid);

由于 create 方式传入了要播放的资源 ID，因此后续就不需要调用 setDataSource 设置资源信息了。

2. 设置要播放的资源文件

MediaPlayer 要播放的资源文件主要有以下三个来源：

- 用户在应用中事先置入的 resource 资源，例如 MediaPlayer.create(this, R.raw.test_media_file_rsid);。
- 存储在 SD 卡或其他文件路径下的媒体文件，例如 mp.setDataSource("/sdcard/test.mp3");。
- 网络上的媒体文件，例如 mp.setDataSource("http://www.xxx.com/test.mp3");。

MediaPlayer 的 setDataSource 一共有四个方法：

- setDataSource (String path)
- setDataSource (FileDescriptor fd)
- setDataSource (Context context, Uri uri)
- setDataSource (FileDescriptor fd, long offset, long length)

3. 对播放器进行控制

Android 通过控制播放器状态方式来控制媒体文件的播放，下面对 Android MediaPlayer 类的重要方法和对应状态进行介绍，见表 4-5。

表 4-5 Android MediaPlayer 类的重要方法和对应状态

方法名	状态改变	说明
prepare()	Initialized→Prepared	同步方式设置播放器进入 prepare 状态
prepareAsync()	Initialized→Preparing→Prepared	异步方式设置播放器进入 prepare 状态
start()	Prepared→Started	启动媒体文件播放
pause()	Started→Paused	暂停媒体文件播放
stop()	Paused→Stopped 或 Started→Stopped	停止媒体文件播放
seekTo()	Prepared、Started、Paused 或 PlaybackCompleted	让播放器从媒体文件指定的位置开始播放
reset()	Idle	设置播放器重新回到初始状态
release()	Idle→End	释放播放器占用的资源
setDataResource()	Idle→Initialized	为播放器设置准备播放的媒体资源

表 4-5 对 Android 中 MediaPlayer 播放器的重要方法和方法调用成功后播放状态的变化进行了说明，接下来还有几点注意事项需要了解：

（1）prepare()和 prepareAsync()提供了同步和异步两种方式设置播放器进入 prepare 状态，如果 MediaPlayer 实例是通过 create()方法创建的，则第一次开始播放前不需要再调用

prepare()了，因为 create()方法里已经默认调用过 prepare()方法。

（2）seekTo()是一个异步方法，也就是说该方法返回时并不意味着定位完成，尤其当播放的媒体文件为远程网络文件时，真正定位完成就会触发 OnSeekCompleteListener 监听器下的 onSeekComplete()方法，因此我们可以调用 setOnSeekCompleteListener(OnSeekCompleteListener)方法来设置监听器，对 seekTo 的结果进行处理。

（3）一旦确定不再使用播放器时，应当尽早调用 release()方法来释放资源。

（4）reset()可以使播放器从 Error 状态中恢复过来，当播放器出现错误时应该考虑调用该方法。

（5）在开发媒体播放器相关功能时需要考虑到播放器可能出现的情况，设置好监听和处理逻辑，以保持播放器的健壮性，除了 OnSeekCompleteListener 监听器，MediaPlayer 还提供了 OnCompletionListener 监听器和 OnErrorListener 监听器来更好地对播放器的工作状态进行监听，供开发者及时处理各种情况：

1）setOnCompletionListener(MediaPlayer.OnCompletionListener listener)用于监听播放的媒体文件是否正常播放完毕。

2）setOnErrorListener(MediaPlayer.OnErrorListener listener)用于监听播放器使用过程中是否出现错误。

4.4 关键知识点解析

4.4.1 图片加载到内存 OOM

在 4.3.4 节中，我们介绍了 Android 中图片压缩的原理和方法，读者已经了解到如何将图片压缩至更小尺寸进行存储或传输。然而还有一个问题没有解决，如果将一张在存储器或网络上的大图片加载至内存中，当应用程序的内存不足以加载这张图片时，会有什么情况发生呢？

答案是显而易见的，作为导致 Android 应用程序崩溃最常见的内存溢出（Out Of Memory，OOM）问题，应用程序内存空间无法满足所加载的图片的要求是一个非常直接的原因。接下来我们也许还要问，如果已经对这张图片进行了压缩处理，能否解决内存溢出的问题呢？

答案是不一定，因为计算机存储、传输图片与加载图片到内存的机制是有区别的，内存中图片的大小不是直接由图片的存储大小来决定的。比如将一个 10KB 大小的 png 格式的图片加载到内存所占用的空间就远不止 10KB 了。那这张图片到底会占用多少内存空间呢？请看下面的计算公式：

图片加载到内存中的大小=图片的像素宽×图片的像素高×该图片一个像素所占的位数/8。

举个例子：一个 1024 像素×1024 像素的图片，每个像素存储占 32 位，那么它的大小就是 1024×1024×32÷8=4MB。通常图片经过压缩处理保存成 jpg、png 格式，它的存储大小可能就只有十几 KB，而一旦将其加载到内存，其所占用的空间将达到若干 MB。这就是为什么我们仅仅加载了一张几十 KB 的图片，却可能出现 OOM 异常的原因。

因此，我们需要考虑对加载到内存的图片进行压缩处理，以确保有足够的可用内存来存储这张图片。这就要从 BitmapFactory 这个类所提供的图片加载入内存的相应方法说起。

BitmapFactory 类提供了多个解析方法（decodeByteArray、decodeFile、decodeResource 等）用于创建 Bitmap 对象，我们应该根据图片的来源选择合适的方法。比如 SD 卡中的图片可以使用 decodeFile 方法，二进制流形式的图片可以使用 decodeStream 方法，资源文件中的图片可以使用 decodeResource 方法。这些方法会尝试为已经构建的 Bitmap 图片分配内存，而此时我们需要关注这些解析方法所共同提供的一个可选的 BitmapFactory.Options 参数，灵活地配置这个参数，将有效降低内存溢出异常的发生率，并提高程序的性能。

通过前面的计算公式可以分析出，当图片加载入内存时其每个像素所占的位数是一定的，如果希望占用更少的内存空间，我们能做的是缩小图片的像素高和像素宽，因此我们需要先得到该图片的实际像素高和像素宽，将 BitmapFactory.Options 参数的 inJustDecodeBounds 属性设置为 true 就可以让解析方法避免为 Bitmap 分配内存，而仅计算出图片的实际宽和高。我们可以通过 BitmapFactory.Options 的 outWidth、outHeight 和 outMimeType 属性在向内存加载图片之前就获取到图片的长宽值和媒体类型，从而根据情况对图片进行压缩。代码如下所示：

```java
BitmapFactory.Options options = new BitmapFactory.Options();
options.inJustDecodeBounds = true;
BitmapFactory.decodeResource(getResources(), R.id.myimage, options);
int imageHeight = options.outHeight;          //图片的实际像素高
int imageWidth = options.outWidth;            //图片的实际像素宽
String imageType = options.outMimeType;       //图片的媒体类型
```

在加载图片时，建议事先检查一下图片的大小，除非能确保这个图片不会导致 OOM 异常。

通过上面的代码我们能得到图片的实际尺寸，下面对图片进行压缩处理。通过设置 BitmapFactory.Options 中 inSampleSize 属性的值就可以对图片的宽高进行等比缩小。例如我们需要加载一张 1024 像素×768 像素的图片，将 inSampleSize 的值设置为 4，就可以把这张图片压缩成 256 像素×192 像素并加载至内存中。原本加载这张图片需要占用 3MB 的内存，压缩后只需要占用 192KB（假设图片是 ARGB_8888 类型，即每个像素点占用 4 个字节）。那么如何计算出合适的 inSampleSize 值呢？我们需要将图片实际的尺寸与在内存中加载时的目标尺寸进行对比计算，得到最合适的 inSampleSize 值，其参考代码如下：

```java
public static int calculateInSampleSize(BitmapFactory.Options options, int reqWidth, int reqHeight) {
    //原始图片的高度和宽度
    final int height = options.outHeight;
    final int width = options.outWidth;
    int inSampleSize = 1;
    if (height > reqHeight || width > reqWidth) {
        //计算出实际像素宽和实际像素高与目标宽高的比率
        final int heightRatio = Math.round((float) height / (float) reqHeight);
        final int widthRatio = Math.round((float) width / (float) reqWidth);
        //选择宽和高中最小的比率作为inSampleSize的值，这样可以保证最终图片的宽和高比例
        //一定都会大于等于目标的宽和高
        inSampleSize = heightRatio < widthRatio ? heightRatio : widthRatio;
    }
    return inSampleSize;
}
```

需要注意的是，使用 calculateInSampleSize 这个方法时，需要先计算出图片的实际宽

高，即上文提到过的将 BitmapFactory.Options 的 inJustDecodeBounds 属性设置为 true，进行第一次解析，然后将 BitmapFactory.Options 连同目标的宽度和高度一起传递到 calculateInSampleSize 方法中，就可以得到合适的 inSampleSize 值了。之后我们需要使用新获取到的 inSampleSize 值再一次解析图片，并把 inJustDecodeBounds 设置为 false，就可以在内存中载入压缩后的图片了。

4.4.2 大量图片的缓存处理

目前，我们已经可以轻而易举地在移动应用的 UI 界面中加载一张图片，但当加载图片的数量不断增多时，情况会变得愈发复杂。例如，在使用 ListView、GridView 或者 ViewPager 等组件时，屏幕上的图片可能会随着用户的手势滑动事件而不断增加，如果不进行有效处理，很容易出现 OOM 错误，直接导致应用程序崩溃。

此时，可能会有读者想到通过适时将已加载过的图片进行回收处理，同时将应用可用内存维持在一个合理可控的范围内。用这种思路来解决问题是非常好的，但我们却无法回避另一些问题的挑战，即在什么条件下触发回收图片？回收哪些图片？如果图片被回收之后，用户又将它重新滑入屏幕，我们是否还要重新加载曾经加载（很多时候需要重新通过网络下载）过的图片？

那么到底有没有一种更好的方法能同时解决上述问题呢？接下来我们将使用 LruCache 缓存技术来处理这个问题。我们可以将图片缓存至 LruCache，由 LruCache 为我们管理图片。LruCache 会自动计算缓存中哪些图片更加常用，哪些图片不太常用，从而在有限内存的条件下在内存中保留更多常用的图片，而当内存吃紧时清理部分最不常用的图片，下面先来看一个例子：

```
private LruCache<String, Bitmap> mMemoryCache;

@Override
protected void onCreate(Bundle savedInstanceState) {
    //获取到可用内存的最大值，使用内存超出这个值会引起OutOfMemory异常
    //LruCache通过构造函数传入缓存值，以KB为单位
    int maxMemory = (int) (Runtime.getRuntime().maxMemory() / 1024);
    //使用最大可用内存值的1/8作为缓存的大小
    int cacheSize = maxMemory / 8;
    mMemoryCache = new LruCache<String, Bitmap>(maxSize) {
        @Override
        protected int sizeOf(String key, Bitmap bitmap) {
            //重写此方法来衡量每张图片的大小，默认返回图片数量
            return bitmap.getByteCount() / 1024;
        }
    };
}

public void addBmpToLruCache(String key, Bitmap bitmap) {
    if (getBmpFromLruCache(key) == null) {
        mMemoryCache.put(key, bitmap);
```

```
        }
    }

    public Bitmap getBmpFromLruCache(String key) {
        return mMemoryCache.get(key);
    }
```

从上面的例子可以看出，在对 LruCache 进行创建的过程中，需要向构造函数中传入 maxSize 参数并实现 sizeOf()方法，其中 cacheSize 参数代表 LruCache 可以使用的缓存大小，以 KB 为单位。如果存储的图片超过这个大小，LruCache 就会采用"最近—最少使用"策略来删除部分缓存数据，直至缓存大小小于 maxSize 的大小。例子中使用了该进程能够从系统获得的最大内存的 1/8 作为 LruCache 的缓存大小。另外，我们重写了 sizeOf()方法，该方法默认返回 LruCache 中放入的图片数量，其作用主要用于定义缓存中每项的大小。当存入一张图片后，LruCache 就会调用 sizeOf()方法计算新加入图片的尺寸，并与已使用的大小进行叠加成为新的已使用缓存大小，如果已使用的缓存大小超过 maxSize 就会对图片进行清理，直至缓存中的图片大小总和小于 maxSize。在 LruCache 构建完毕后，我们又新建了两个方法——addBmpToLruCache (String key, Bitmap bitmap)和 getBmpFromLruCache(String key)，用于向 LruCache 中增加图片和从 LruCache 中获取图片，其使用方式实际上与 HashMap 是一致的，都是通过 put 方法存入，get 方法获取。

通过上述的实例，我们已经掌握 LruCache 的基础用法。这里需要注意的是，我们需要判断 getBmpFromLruCache(String key)方法返回值是否为空，如果为空，则需要重新将图片加入 LruCache。在实际使用中，LruCache 支持对曾经清理掉的图片的重建，用户还可以自己通过 LruCache 的 create()方法来对其进行相关实现。

4.5 问题与讨论

1．如何读取手机图库？
2．图像特效的处理方法都有哪些？
3．如何使用 LruCache 提升应用程序的性能和效率？

项目 5　学习监督器

本项目从学习监督器入手,为读者呈现一套较完整的基于 Activity 和 Service 组件的项目建设流程。本项目着重对 Android 中 Service 组件的基本用法进行介绍,继而对 Service 组件的进阶用法进行讲解。

项目需求描述如下:

(1)通过使用 Service 组件在后台监控网络状态,用户在学习期间一旦使用手机网络,学习监督应用便会进行警告。

(2)用户可以设置学习的时间段,而一旦时间被设置后,则不能轻易取消,如果要退出监督模式,需要先接受应用的惩罚。

(1)掌握启动(start)Service 的方法。
(2)掌握绑定(bind)Service 的方法。
(3)掌握 Service 与 Activity 直接交互通信的方法。
(4)掌握在前台运行服务。
(5)学会管理服务的生命周期。
(6)使用 AIDL 进行进程间通信。
(7)掌握 MediaPlayer 的用法。

5.1　总体设计

5.1.1　总体分析

根据项目需求进行分析,学习监督器本身的功能比较简单,每次启动时,首先展示一个欢迎界面,短暂停留后将进入应用主界面。主界面中可以设定起始时间和结束时间,在此时间范围内,学习监督器将对用户进行上网行为监督,一旦用户打开网络链接,无论是 4G 网络还是 Wi-Fi 网络,学习监督器都会发现并且对用户进行警告,警告包括巨大声音的音乐和持续不停的振动,直至用户关闭网络,警告行为才停止。

5.1.2　功能模块框图

根据总体分析结果可以总结出功能模块框图,本学习监督器主要包括时间设置、监督提醒和后台监控三个模块，功能模块框图如图 5-1 所示。

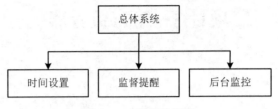

图 5-1　功能模块框图

后台监控服务具体功能如图 5-2 所示，其对网络状态变化进行监控，一旦发现网络状态产生变化则进行相应的事件处理，主要有如下两种场景：

（1）在监督时间范围内发现用户开启了 Wi-Fi 或 4G 网络，后台监控服务则通知系统启动监督提醒界面，进行相应预设好的提醒。

（2）当监督提醒的大声音播放音乐和持续振动发生时，如果用户关闭网络，后台监控服务则通知系统停止音乐播放和持续振动。

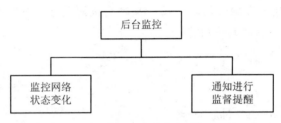

图 5-2　后台监控方法

监督提醒模块具体功能如图 5-3 所示，当其得到系统通知，需要进行提醒时，开启提醒界面，进行音乐播放并触发手机持续振动。

图 5-3　监督提醒模块功能图

5.1.3　系统流程图

根据总体分析结果及功能模块框图梳理出系统的主要流程，如图 5-4 所示。

5.1.4　界面设计

接下来开始考虑界面的设计，根据上述对需求的分析可以确定，至少有两个主要界面，即时间设置界面和监督提醒界面，同时考虑到应用的完整性，我们还需要设计欢迎界面和帮助说明界面，各界面设计效果如图 5-5 至图 5-8 所示。

项目 5 学习监督器

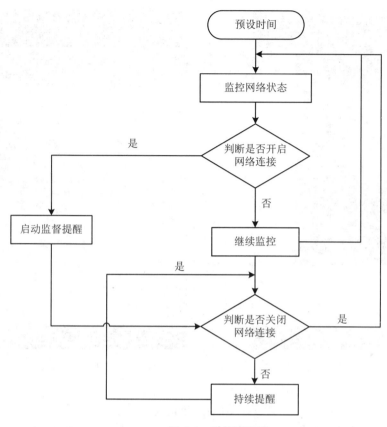

图 5-4 系统流程图

图 5-5 欢迎界面设计图

图 5-6 时间设置界面设计图

图 5-7　监督提醒界面设计图　　　　图 5-8　帮助说明界面设计图

5.2　详细设计

5.2.1　系统包及其资源规划

对应界面设计图，本系统至少需要 4 个 Activity 类，欢迎界面对应 SplashActivity，时间设置界面对应 MainActivity，监督提醒界面对应 PunishActivity，帮助说明界面对应 GuideActivity。本项目我们将学习使用 Android Studio 来进行创建和开发，包及其资源结构如图 5-9 所示。

5.2.2　时间设置 Activity 设计

时间设置 Activity 作为应用的主 Activity，负责与用户交互的主要环节，包括接收用户设置的时间、保存时间、初始化后台服务等。

设置时间可以考虑使用 Android 原生 TimePicker 控件，也可以使用包含 TimePicker 的 TimePickerDialog 弹出框。

保存时间的设置信息可以使用 Android 的 SharedPreferences 进行简单文本信息的存储与读取。

初始化后台服务有两种方式：启动服务（startService）和绑定服务（bindService）。这也是本项目学习的重点，我们会在 5.3 节中着重讲解。

项目 5　学习监督器　95

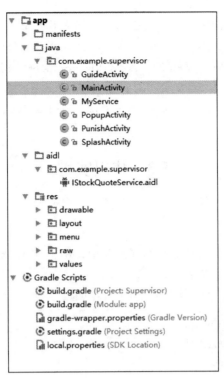

图 5-9　包及其资源结构

5.2.3　后台服务设计

Android 系统允许用户自定义服务运行在后台，这就给应用程序的开发增加了很多便利。实现后台服务的组件叫做 Service，作为可以在后台执行长时间运行操作而不提供用户界面的应用组件，Service 一般由其他应用组件（Activity、Service、BroadcastReceiver）启动，即使应用被用户切换到后台，服务仍将在后台继续运行。

5.2.2 节提到过，服务分为启动服务（startService）和绑定服务（bindService）。前者是由应用组件调用startService()方法发起的，比如在一个 Activity 的 onCreate()方法中调用 startService()方法，此时 Service 将处于启动状态。一旦启动，服务即可在后台无限期运行（目前市场上很多品牌的 Android 手机为了有效节省电量，会主动停止后台服务），即使启动服务的组件已被销毁也不受影响。已启动的服务通常是执行单一操作，而且不会将结果返回给调用方。例如，使用服务从网络下载或上传文件，当操作完成后，服务可以自行停止运行。

不同于启动服务，绑定服务是由应用组件调用bindService()方法发起的，此时服务将处于绑定状态，只有当服务与某个组件绑定成功时，绑定服务才能运行，这种服务提供了一个 C/S 接口，允许绑定服务的组件与服务进行交互、发送请求、获取结果。例如，播放音乐、执行网络或文件 I/O 操作、与 ContentProvider 交互时，都需要绑定服务的组件与启动后的服务进行数据交互。多个组件可以同时绑定到该服务，但全部取消绑定后，该服务会被销毁。

对于 Service 组件，有一点读者要注意，Service 组件在其托管进程的主线程中运行，和我们常说的 Activity UI 主线程是同一线程，如不特别指定，系统不会为 Service 组件创建独立的线程，也不在单独的进程中运行。这意味着，如果 Service 中需要执行任何 CPU 密集型工作或阻塞操作，例如音乐播放或进行网络交互，则应在 Service 内创建新线程来完成这项工作。通过使用单独的线程，可以降低发生 ANR（应用无响应）错误的风险，而应用的主线程仍可继续专注于运行用户与 Activity 之间的交互工作。

为了让读者更全面地掌握 Service 的使用方法，接下来我们分别使用两种模式对学习监督器加以实现。

5.3 代码实现

5.3.1 显示界面布局

系统主界面使用 LinearLayout 进行布局，其中使用了 TimePicker 控件用于设置监督的起始时间和结束时间，主要代码如下：

```xml
<LinearLayout xmlns:android="http://schemas.android.com/apk/res/android"
    android:layout_width="fill_parent"
    android:layout_height="fill_parent"
    android:background="@drawable/main_bg"
    android:orientation="vertical" >
 <Button
        android:layout_width="wrap_content"
        android:layout_height="wrap_content"
        android:text="帮助说明"
        android:id="@+id/userButton" />
   <LinearLayout
      android:layout_width="wrap_content"
      android:layout_height="wrap_content"
      android:orientation="horizontal" >
      <TextView android:id="@+id/start" android:layout_height="wrap_content"
            android:layout_width="wrap_content"
             android:textSize="20sp"
             android:text="开始时间： " >
</TextView>
        <TimePicker
            android:id="@+id/mTimPicker"
            android:layout_width="wrap_content"
            android:layout_height="wrap_content"/>
   </LinearLayout>

       <LinearLayout
      android:layout_width="wrap_content"
      android:layout_height="wrap_content"
      android:orientation="horizontal" >
      <TextView android:id="@+id/end" android:layout_height="wrap_content"
            android:layout_width="wrap_content"
```

```
                    android:textSize="20sp"
                    android:text="结束时间: " >
        </TextView>
            <TimePicker
                android:id="@+id/mTimPicker2"
                android:layout_width="fill_parent"
                android:layout_height="wrap_content"/>
        </LinearLayout>
         <Button
            android:layout_width="wrap_content"
            android:layout_height="wrap_content"
            android:text="设    置"
            android:id="@+id/button" />
</LinearLayout>
```

其他几个界面的布局相对简单，这里不再赘述，读者可以直接参考工程代码中的 guide_layout.xml、punish_layout.xml、splash_layout.xml 三个布局。

5.3.2 构建一个服务

要构建服务就必须创建Service的子类，继承 Service 类，同时重写部分回调方法，用来对服务生命周期中的关键事件进行处理。下面介绍需要重写的几个重要回调方法。

（1）onStartCommand()。当另一个组件（如 Activity）通过调用startService()请求启动服务时，系统将调用此方法。一旦执行此方法，服务即会启动并可在后台无限期运行。如果实现了此方法，则在服务工作完成后，需要通过调用stopSelf()或stopService()来停止服务。相反，如果我们只想使用 bindService()方法来绑定 Service，则无需实现此方法。

（2）onBind()。当另一个组件（如 Activity）想通过调用bindService()与 Service 绑定（例如执行 RPC）时，系统将调用此方法。在此方法的实现中，必须通过返回 IBinder 提供一个接口，供客户端与服务进行通信。在 bindService()的模式下，这个方法是务必要实现的，但如果不允许任何组件绑定这个服务，则可以返回 null。

（3）onCreate()。首次创建 Service 时，系统将调用此方法来执行一次性设置程序（在调用 onStart-Command()或 onBind()之前）。如果 Service 已在运行，则不会调用此方法。

（4）onDestroy()。当服务不再使用且将被销毁时，系统将调用此方法。服务应该实现此方法来清理所有资源，如线程、注册的监听器、接收器等。这是服务接收的最后一个调用。

由上述内容可知，服务的创建可以有两种实现方式，为了让读者更加清楚这两种方式的用法和区别，接下来分别使用不同的方式对其进行实现。不过无论使用哪种服务方式，构建服务的基本步骤都是类似的，我们需要创建一个继承 Service 类的自定义类，命名为 MonitorService.java，同时重写 onCreate、onDestroy、onStartCommand、onBind 方法，代码如下：

```
public class MonitorService extends Service
{
    static final String TAG="MonitorService";
    @Override
    public void onCreate() {
```

```java
        super.onCreate();
    }
    @Override
    public IBinder onBind(Intent arg0)
    {
        //仅通过startService()方式启动服务而不需要绑定服务时，onBind()方法可以返回null
        return   null;
    }
    @Override
    public void onDestroy()
    {
        Log.e(TAG, "Release MonitorService");
        super.onDestroy();
    }

    @Override
    public int onStartCommand(Intent intent, int flags, int startId) {

    }
}

public class MonitorService extends Service
{
    static final String TAG="MonitorService";
    @Override
    public void onCreate() {
        super.onCreate();
    }

    @Override
    public IBinder onBind(Intent arg0)
    {
        //返回AIDL实现
        return new MyServiceImpl();
    }
    @Override
    public void onDestroy()
    {
        Log.e(TAG, "Release MonitorService");
        super.onDestroy();
    }

    @Override
    public int onStartCommand(Intent intent, int flags, int startId) {
        return super.onStartCommand(intent, flags, startId);
    }
}
```

如此对 Service 类进行了初步定义。在具体丰富其业务代码之前,我们需要在 AndroidManifest.xml 文件中对服务进行声明,需要添加 <service> 元素作为 <application> 元素的子元素。代码如下:

```
<service android:name="com.example.supervisor.MonitorService" android:exported="false" >
</service>
```

android:name 属性是唯一必需的属性,用于指定服务的类名。为了确保应用的安全性,请始终使用显式 Intent 启动或绑定 Service,且不要为服务声明 Intent 过滤器。

此外,还可以通过添加 android:exported 属性并将其设置为 false 来确保服务仅适用于您的应用。这可以有效阻止其他应用启动您的服务。

5.3.3 创建启动服务

接下来我们通过创建启动服务的方式来实现学习监督器的后台服务功能,启动服务由另一个组件通过调用startService()来启动,这会导致调用服务的onStartCommand()方法。

服务启动之后,其生命周期即独立于启动它的组件,并且可以在后台无限期地运行,即使启动服务的组件已被销毁也不受影响。因此,服务应通过调用 stopSelf() 结束工作来自行停止运行,或者由另一个组件通过调用 stopService()来停止它。

应用组件(如 Activity)可以通过调用 startService()方法并传递 Intent 对象(指定服务并包含待使用服务的所有数据)来启动服务。服务通过 onStartCommand()方法接收此 Intent。

因此,我们需要实现 onStartCommand()方法,代码如下:

```
@Override
public int onStartCommand(Intent intent, int flags, int startId) {
    Log.e(TAG,"onStartCommand() has been invoked!") ;
    monitorNetwork();
    return super.onStartCommand(intent, flags, startId);
}
```

在 MainActivity 中创建 startMonitorService()方法,通过创建 Intent 和调用 startService(intent)方法来启动 MonitorService,代码如下:

```
private void startMonitorService() {
Log.e(TAG, "start Service！ ");
    Intent intent = new Intent(this, MonitorService.class);
    startService(intent);
}
```

由于在 Service 中,每次调用 startService()方法时 onStartCommand()都会被调用,因此我们可以通过这种机制用 Activity 控制 Service,也可以用这种方式在 Intent 中加入参数为 Service 传参。

接下来分析具体需要在什么时候调用 startMonitorService()方法。

一是 Activity 在启动时需要启动 Service,我们首先调用 startMonitorService()方法来启动 MonitorService,继而对 Button 和 TimePicker 控件进行初始化。代码如下:

```
@Override
public void onCreate(Bundle savedInstanceState) {
    super.onCreate(savedInstanceState);
    setContentView(R.layout.main_layout);
```

```java
        startMonitorService();        //启动服务
        Button btnCall = (Button) findViewById(R.id.button);
        Button userButton = (Button) findViewById(R.id.userButton);
        //加入时间设置
        mTimePicker = (TimePicker) findViewById(R.id.mTimPicker);
        mTimePicker2 = (TimePicker) findViewById(R.id.mTimPicker2);
        //用于保持上一次数据
        sharedPreferences = this.getSharedPreferences(mSettingFlag, MODE_WORLD_READABLE);
        editor = sharedPreferences.edit();
        //获取上一次设置的数据
        startHour = sharedPreferences.getInt("startHour", 8);
        startMinite = sharedPreferences.getInt("startMinite", 0);
        endHour = sharedPreferences.getInt("endHour", 23);
        endMinute = sharedPreferences.getInt("endMinute", 0);
        mTimePicker.setIs24HourView(true);                //是否显示24小时制，默认为false
        mTimePicker.setCurrentHour(startHour);            //设置当前小时
        mTimePicker.setCurrentMinute(startMinite);        //设置当前分钟
        mTimePicker.setOnTimeChangedListener(new OnTimeChangedListener() {
            public void onTimeChanged(TimePicker view, int hourOfDay, int minute) {
                editor.putInt("startHour", hourOfDay)
                    .putInt("startMinite", minute)
                        .commit();
                startHour = hourOfDay;
                startMinite = minute;
            }
        });
        //设置mTimePicker2为24小时制
        mTimePicker2.setIs24HourView(true);
        //设置mTimePicker2初始值为5
        mTimePicker2.setCurrentHour(endHour);            //设置当前小时
        mTimePicker2.setCurrentMinute(endMinute);
        //设置mTimePicker2时间改变事件处理
        mTimePicker2.setOnTimeChangedListener(new OnTimeChangedListener() {
            public void onTimeChanged(TimePicker view, int hourOfDay, int minute) {
                endHour = hourOfDay;
                endMinute = minute;
                editor.putInt("endHour", hourOfDay)
                    .putInt("endMinute", minute)
                        .commit();
            }
        });
        ...
    }
```

二是当用户对时间进行修改后，点击"设置"按钮时需要通知 Service 更新状态，代码如下：

```java
        btnCall.setOnClickListener(new OnClickListener() {
```

```java
        @Override
        public void onClick(View v) {
            startMonitorService();
        }
    });
```

通过以上两个环节,确保 MonitorService 每次被启动的同时 Service 中的 onStartCommand() 方法将被同时调用。而在 startMonitorService()方法第一次被调用时,MonitorService 中的 onCreate()方法会进行初始化,只要 MonitorService 不被销毁,onCreate()方法就不会被再次调用,代码如下:

```java
private Vibrator vibrator;
private MediaPlayer mp;
private static final String tag="tag";
private ConnectivityManager connectivityManager;
private NetworkInfo info;
//时间变量
private int startHour;
private int startMinite;
private int maxVolumn =10;
private int endHour;
private int endMinute;
private int countTime1;
private int countTime2;
MediaPlayer.OnPreparedListener preparedListener;
private SharedPreferences sharedPreferences;
static final String TAG="MonitorService";
//...
@Override
public void onCreate() {
    super.onCreate();
    connectivityManager = (ConnectivityManager)getSystemService(Context.CONNECTIVITY_SERVICE);
        mAudioManager = (AudioManager) getSystemService(Context.AUDIO_SERVICE);
    mp=MediaPlayer.create(MonitorService.this, R.raw.jikechufa);
    mp.setLooping(true);
    preparedListener = new MediaPlayer.OnPreparedListener() {
        @Override
        public void onPrepared(MediaPlayer mp) {
            mp.start();
        }
    };
    IntentFilter mFilter = new IntentFilter();
    mFilter.addAction(ConnectivityManager.CONNECTIVITY_ACTION);
    registerReceiver(mReceiver, mFilter);
        //获取振动服务
        vibrator = (Vibrator)getSystemService(Context.VIBRATOR_SERVICE);
            //用于保持上一次数据
            sharedPreferences = this.getSharedPreferences(MainActivity.mSettingFlag,MODE_WORLD_READABLE);
```

```
        monitorNetwork();
    }
```

5.3.4 监控网络变化

当系统网络发生变化时（如 Wi-Fi 切换、4G 切换等），系统会向应用程序发送广播，应用程序可以通过注册广播接收器来接收通知，我们可以在 MonitorService 中注册一个专门用来接收系统网络变化的广播接收器，代码如下：

```
//注册广播接收器进行网络状态的监听
private BroadcastReceiver mReceiver = new BroadcastReceiver() {
        @Override
        public void onReceive(Context context, Intent intent) {
            String action = intent.getAction();
            if (action.equals(ConnectivityManager.CONNECTIVITY_ACTION)) {
                Log.e(tag, "nwtwork changed");
                monitorNetwork();
            }
        }
};
```

细心的读者可能会发现，onStartCommand()方法和本节的 onReceive()方法中都存在一个 monitorNetwork()方法，这个方法便是监控到网络变化后具体的业务处理方法，代码如下：

```
private void monitorNetwork(){
    info = connectivityManager.getActiveNetworkInfo();
    if(info != null && info.isAvailable()) {
        String name = info.getTypeName();
        Log.e(tag, "Network Name:" + name);
            //获取时间
            if(compareTime()){

                if(null!=mp){
                    play();
                }
                vibrator.vibrate(new long[] { 100,1000,100,1000 }, 0);

            }else{
                Log.e(tag, "time not in range");
                if(null!=vibrator){    vibrator.cancel();    }
                if(null!=mp&&mp.isPlaying()){
                    mp.stop();
                }
            }
    } else {
        Log.e(tag, "No available network");
        if(null!=vibrator){    vibrator.cancel();    }
        if(null!=mp&&mp.isPlaying()){
            mp.stop();
```

                }
            }
        }

此段代码中涉及 MediaPlayer 的使用，我们将其使用方法进行了简单封装，代码如下：
```
public void play() {
        try {
            if(mp !=null){
                mp.stop();
            }
            mp.setOnPreparedListener(preparedListener);
            mp.prepareAsync();
        } catch (IllegalStateException e) {
            e.printStackTrace();
        }
        //为了使监控提醒更具效果，在惩罚时调高手机播放的音量
        int currentVolumn = mAudioManager.getStreamVolume(AudioManager.STREAM_MUSIC);
        if (currentVolumn < maxVolumn) {
            mAudioManager.setStreamVolume(AudioManager.STREAM_MUSIC, maxVolumn, 0);
        }
}
```

5.3.5　时间比较

当用户进行监控时间范围设置后，MonitorService 将进行当前时间和用户设置时间的比较，如果当前时间在用户设置时间的范围内，则返回 true，否则返回 false，代码如下：
```
private boolean compareTime(){
    final Calendar c = Calendar.getInstance();
    int currentHour = c.get(Calendar.HOUR_OF_DAY);        //获取当前的小时数
    int currentMinute = c.get(Calendar.MINUTE);           //获取当前的分钟数
    startHour = sharedPreferences.getInt("startHour", 8);
    startMinite = sharedPreferences.getInt("startMinite", 0);
    endHour = sharedPreferences.getInt("endHour", 23);
    endMinute = sharedPreferences.getInt("endMinute", 0);

    Log.e(TAG,"currentHour:"+currentHour) ;
    Log.e(TAG,"currentMinute:"+currentMinute) ;
    countTime1=startHour*60+ startMinite;
    countTime2= endHour *60+ endMinute;
    Log.e(TAG,"get countTime1:"+countTime1) ;
    Log.e(TAG,"get countTime2:"+countTime2) ;

    int nowCountTime=currentHour*60+currentMinute;
    Log.e(TAG,"get currentTime:"+nowCountTime) ;
     if(countTime1<countTime2){
        if(nowCountTime>=countTime1&&nowCountTime<=countTime2){
```

```
            Log.e(TAG,"Right！！！！！！！！ ")；
            return true;
       }
     }else{
       if(nowCountTime>=countTime1&&nowCountTime<=60*24||nowCountTime>=
           0&&nowCountTime<=countTime1)
       {
            Log.e(TAG,"right！！！！！！！！ ")；
            return true;
       }
     }

    return false;
}
```

5.3.6 创建绑定的服务

在前面几节中我们通过 startService 启动了服务，并多次通过 startService 对 MonitorService 中的 monitorNetwork()方法进行了间接调用（通过 onStartCommand()方法回调）。接下来我们学习如何创建一个绑定的服务，并通过直接调用 Service 实例来操作 Service 中的方法。

下面是具体的操作方法。

（1）首先在 MonitorService 类中创建一个可满足下列任一项的Binder 实例：
- 包含可以被客户端调用的公共方法。
- 返回值为当前 Service 实例，其中包含客户端可调用的公共方法。
- 返回由服务承载的其他类的实例，其中包含客户端可调用的公共方法。

（2）从onBind()回调方法返回此 Binder 实例。

（3）在客户端中，从onServiceConnected()回调方法接收 Binder，并使用提供的方法调用绑定服务。

我们需要在 MonitorService 中创建如下代码：

```
private final IBinder mBinder = new LocalBinder();

public class LocalBinder extends Binder {
        public MonitorService getService() {
            return MonitorService.this;
        }
}
@Override
public IBinder onBind(Intent intent) {
    return mBinder;
}
```

如此便可以在 MainActivity 中绑定 MonitorService，进而获得 MonitorService 的实例，直接调用 MonitorService 中的方法。

首先构建 ServiceConnection 的实例，在 onServiceConnected 方法中将 IBinder 转换为

MonitorService 的实例，代码如下：
```
private ServiceConnection mConnection = new ServiceConnection() {

    @Override
    public void onServiceConnected(ComponentName className,IBinder service) {
        MonitorService.LocalBinder binder = ( MonitorService.LocalBinder) service;
        mService =   binder.getService();
    }

    @Override
    public void onServiceDisconnected(ComponentName arg0) {

    }
};
```
接下来开始绑定服务，代码如下：
```
MonitorService mService;
private void bindMonitorService() {
    Intent intent = new Intent(this, MonitorService.class);
    bindService(intent, mConnection, Context.BIND_AUTO_CREATE);
}
```
在 MainActivity 的 onCreate()方法中注释掉之前的 startMonitorService()方法，调用 bindMonitorService()方法，代码如下：
```
public void onCreate(Bundle savedInstanceState) {
    super.onCreate(savedInstanceState);
    setContentView(R.layout.main_layout);
    //startMonitorService();      //启动服务
    bindMonitorService();         //绑定服务
    Button btnCall = (Button) findViewById(R.id.button);
    Button userButton = (Button) findViewById(R.id.userButton);
    //...
}
```
同样，在 MainActivity 的 btnCall 按钮点击事件中注释掉之前的 startMonitorService()方法，直接通过 MonitorService 调用 monitorNetwork()方法，代码如下：
```
btnCall.setOnClickListener(new OnClickListener() {
    @Override
    public void onClick(View v) {
        //startMonitorService();
        mService.monitorNetwork();
    }
});
```
如此便完成了服务的绑定，实现了与启动服务同样的功能。

5.3.7 使用 Activity 作为 Dialog

细心的读者可能会发现，我们的工程中还有一个 PopupActivity 类与 popup_layout.xml 布局，这个 Activity 和布局具体是做什么的呢？其实笔者在这里是希望告诉读者一种弹出界面的技巧，在 Android 中，弹出界面不仅可以使用 Dialog 相关的类，如 AlertDialog，还可以将 Activity 设置为 Dialog 的样式来使用，下面介绍具体实现的方法。Activity 和布局本身的创建并没有什么特别，要点在于 AndroidManifest.xml 清单文件中 android:theme 节点的设置，需要对该节点设置合适的样式，设置方法如下：

```
<activity android:name="com.example.supervisor.PopupActivity"
    android:label="Dialog"
    android:theme="@style/theme01">
</activity>
```

其中引用的 theme01 需要在 values/styles.xml 中进行定义，这个自定义的样式需要继承 Android 系统的对话框样式 android:style/Theme.Dialog，定义方法如下：

```
<style name="theme01" parent="android:style/Theme.Dialog"></style>
```

通过这种方式便可以创建 Dialog 效果的 Activity。

5.4 关键知识点解析

5.4.1 在前台运行服务

在 Android 的组件中，Service 的优先级比 Activity 低。因此，当内存资源紧张时，系统会优先选择 Service 进行清除。当创建的服务不想轻易被系统清除时，我们需要将 Service 设置为前台服务。默认状态下创建的 Service 均为后台服务，我们可以在 onStartCommand()方法中调用 startForeground(ID, new Notification())，将服务设置为前台服务。从 startForeground()方法需要传入的两个参数可以看出，第一个参数为唯一标识，第二个参数为状态栏的 Notification。来看下面的例子：

```
Notification notification = new Notification(R.drawable.icon, getText(R.string.ticker_text),
System.currentTimeMillis());
Intent notificationIntent = new Intent(this, ExampleActivity.class);
PendingIntent pendingIntent = PendingIntent.getActivity(this, 0, notificationIntent, 0);

notification.setLatestEventInfo(this, getText(R.string.notification_title),
getText(R.string.notification_message), pendingIntent);

startForeground(ONGOING_NOTIFICATION_ID, notification);
```

注意：提供给 startForeground()的整型 ID 不得为 0。

要从前台移除服务，请调用 stopForeground()。此方法采用一个布尔值，指示是否也移除状态栏通知。此方法不会停止服务，但是如果服务在前台运行时将其停止，则通知也会被移除。

5.4.2 服务的生命周期

虽然服务的生命周期比 Activity 的生命周期简单，但关注如何创建和销毁服务反而更加重要，因为服务更容易被系统在未通知用户的情况下销毁。一旦重要的服务被意外停止或销毁，将影响应用程序的基本功能，因此深入了解服务的生命周期非常重要。

服务生命周期（从创建到销毁）可以遵循两条不同的路径，图 5-10 所示是谷歌官方文档提供的服务的生命周期。

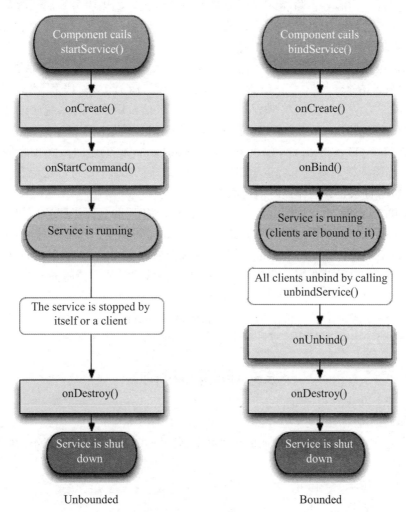

图 5-10　服务的生命周期

图 5-10 中左侧的流程显示了使用startService()所创建的服务的生命周期，右侧的流程显示了使用bindService()所创建的服务的生命周期。从图中可以清楚地看出，服务的整个生命周期从调用onCreate()开始，至onDestroy()返回时结束。与 Activity 类似，服务也在onCreate()中完成初始设置，并在onDestroy()中释放所有剩余资源。无论服务是通过startService()还是bindService()创建，都会调用onCreate()和 onDestroy()方法。

服务的有效生命周期从onStartCommand()或onBind()方法被调用开始。每种方法均可通过Intent对象传递数据到startService()或bindService()方法中。

5.4.3　避免系统回收服务

仅当内存过低且必须回收系统资源以供具有用户焦点的 Activity 使用时，Android 系统才会强制停止服务。如果将服务绑定到具有用户焦点的 Activity，则系统不太可能终止该服务。如果不希望系统回收服务，可以将服务声明为在前台运行，这样一般情况下系统不会强行终止该服务。如果服务已启动并要长时间运行，则系统会随着时间的推移降低服务在后台任务列表中的位置，而服务也将随之变得非常容易被终止。如果服务是启动服务，则必须将其设计为能够妥善处理系统对它的重启。如果系统终止服务，那么一旦资源变得再次可用，系统便会重启服务，不过这还取决于从onStartCommand()返回的值。

由于我们可以自己处理对onStartCommand()的每个调用，因此可以同时执行多个请求，甚至可以为每个请求创建一个新线程，然后立即运行这些线程。要注意的是，onStartCommand() 方法必须返回整型数。整型数是一个值，用于描述系统应该如何在服务终止的情况下继续运行服务，从onStartCommand()返回的值必须是以下常量之一：

（1）START_NOT_STICKY。如果系统在onStartCommand()返回后终止服务，则除非有挂起 Intent 要传递，否则系统不会重建服务。这是最安全的选项，可以避免在不必要时和应用能够轻松重启所有未完成的作业时运行服务。

（2）START_STICKY。如果系统在onStartCommand()返回后终止服务，则会重建服务并调用onStartCommand()，但不会重新传递最后一个 Intent。相反，除非有挂起 Intent 要启动服务（在这种情况下将传递这些 Intent），否则系统会通过空 Intent 调用onStartCommand()。这适用于不执行命令但无限期运行并等待作业的媒体播放器（或类似服务）。

（3）START_REDELIVER_INTENT。如果系统在onStartCommand()返回后终止服务，则会重建服务，并通过传递给服务的最后一个 Intent 调用onStartCommand()。任何挂起 Intent 均依次传递。这适用于主动执行应该立即恢复的作业（例如下载文件）的服务。

5.5　问题与讨论

1．程序如何适应不同的分辨率？
2．按钮有几种不同的状态？如何通过配置 XML 格式布局文件实现在不同状态下显示不同的背景图片？

项目 6 简易网络音乐播放器

在企业级 Android 应用开发中,互联网是极大多数应用获取数据的重要途径,而在众多网络协议中,在移动互联网应用中使用更广泛的是 HTTP 协议。因此,熟练掌握基于 HTTP 协议的网络交互技能是 Android 开发者需要把握的核心问题。学习网络交互相关技术需要首先解决如下两个问题:

(1)由于 Android 系统禁止在 UI 主线程中进行网络交互等耗时操作,因此网络交互通常需要在工作线程中进行,这就需要掌握某些线程间消息传递的机制,为 UI 主线程与工作线程直接建立信息交互的渠道。

(2)数据在网络间传递一般采用 JSON、XML 等格式,在这个过程中需要学习数据解析的相关技术。

项目需求描述如下:

(1)从网络接口获取音频文件列表数据,解析后生成音频列表并显示。

(2)实现音频文件的下载和播放。

(1)学习使用 AsyncTask 在后台线程运行的同时进行 UI 界面刷新。
(2)学习使用和读取资产文件夹(assets)中的文件。
(3)学习利用 HttpURLConnection 工具进行网络通信。
(4)学习将网络文件下载到手机存储。
(5)本地文件缓存机制初涉。
(6)使用 DocumentBuilder 解析 XML。
(7)掌握媒体播放器 MediaPlayer 类的使用。
(8)了解 Service 组件的基本使用方式。
(9)了解 BroadcastReceiver 组件的基本使用方式。
(10)掌握 Activity 与 Service 通信的基本方法。
(11)掌握 Service 通知 Activity 并刷新 UI 界面的方法。

6.1 总体设计

6.1.1 总体分析

手机铃声应实现以下功能:铃声列表显示、铃声文件下载到本地、铃声播放试听,播放铃声包括在线播放和下载播放。

整个程序除总体模块外，主要分为基础架构模块、用户界面模块、数据管理与控制模块和网络通信模块四大部分。在整个系统中，基础架构模块提供各项基础功能供其他模块调用，总体模块控制系统的生命周期，用户界面模块负责界面显示，数据管理与控制模块主要提供数据管理功能，网络通信模块负责与服务器通信。

6.1.2 功能模块框图

根据总体分析结果可以总结出功能模块框图，如图6-1所示。

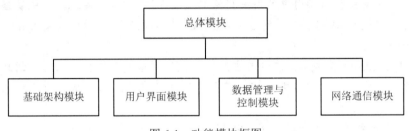

图6-1 功能模块框图

总体模块的作用主要是生成应用程序的主类，控制应用程序的生命周期；基础架构模块主要提供程序架构、所有Activity公用的父类、所有Activity公用的方法，包括自定义风格对话框、自定义提示框等功能；数据管理与控制模块主要提供数据获取、数据解析、数据组织和数据缓存功能；用户界面模块包括铃声列表、铃声播放等功能；网络通信模块主要负责从服务端获取铃声文件数据信息，数据从指定网络地址服务器通过HTTP方式下载获取。

6.1.3 系统流程图

根据总体分析结果及功能模块框图梳理出系统启动的主要流程，如图6-2所示。

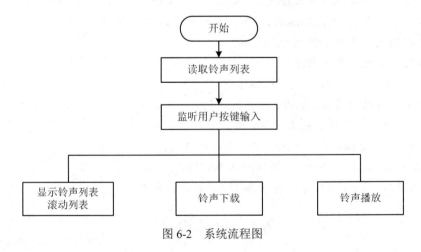

图6-2 系统流程图

6.1.4 界面设计

在系统总体分析及功能模块划分清楚后，即可开始考虑界面的设计。本应用是显示铃声的应用,支持铃声文件下载到本地和铃声播放试听。

根据程序功能需求,可以规划出软件的主要界面如下:

(1)铃声列表:启动应用程序后显示铃声列表,供用户查看铃声。

(2)铃声下载按钮:用户在列表中选中铃声后,点击"下载"按钮,应用可将铃声下载到本地。

(3)铃声播放按钮:用户在列表中选中铃声后,可以播放或暂停该铃声。

系统主界面如图 6-3 所示。

图 6-3　系统主界面

从图 6-3 中可以很直观地看到,界面划分了两个区域:铃声列表区和铃声播放区。

铃声列表区用于显示铃声名称、铃声位置和铃声类别,铃声播放区用于显示当前选中的铃声名称、铃声类别、播放或暂停按钮、下载按钮。

6.2　详细设计

6.2.1　模块描述

在系统总体分析及界面布局设计完成后,主要工作就转入对各个功能模块的详细设计阶段。

1. 基础架构模块详细设计

基础架构模块主要提供程序架构、所有 Activity 公用的父类、所有 Activity 公用的方法,包括自定义风格对话框、自定义提示框等功能。

基础架构模块功能如图 6-4 所示。

图 6-4　基础架构模块功能图

2. 用户界面模块详细设计

用户界面模块的主要任务是显示铃声列表、功能按钮、实现与用户的交互，即当用户点击按键或者屏幕的时候，监听器会去调用相应的处理办法或其他相应的处理模块。

本模块包括铃声列表显示、选中的铃声信息显示、播放和暂停按钮等功能。

用户界面模块功能如图 6-5 所示。

图 6-5　用户界面模块功能图

3. 数据管理与控制模块详细设计

数据管理与控制模块主要提供数据获取、数据解析、数据组织和数据缓存功能。

数据管理与控制模块和用户界面模块可以调用基础架构模块的一些通用方法。数据管理与控制模块为用户界面模块提供数据，同时可以接收并保存用户界面模块产生的数据。

数据管理与控制模块功能如图 6-6 所示。

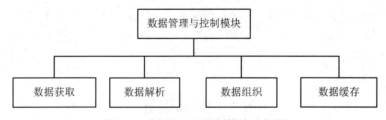

图 6-6　数据管理与控制模块功能图

4. 网络通信模块详细设计

网络通信模块根据用户界面的需求调用访问服务器，接收服务器返回的数据并解析，同时显示到用户界面。

本模块包括发送网络请求、接收网络应答、网络数据解析等功能。

网络通信模块功能如图 6-7 所示。

图 6-7 网络通信模块功能图

6.2.2 系统包及其资源规划

1. 文件结构

根据系统功能设计，本系统封装一个基础的 Activity 类，加载各个子页的通用控件，并提供一些基础的实现方法，例如设置进度条、标题等常用的方法，程序中的 Activity 都可继承此基类，继承后就可直接使用基类中封装的基础方法。

系统使用一个 Activity，用于显示铃声列表、铃声下载和播放操作按钮。包及其资源结构如图 6-8 所示。

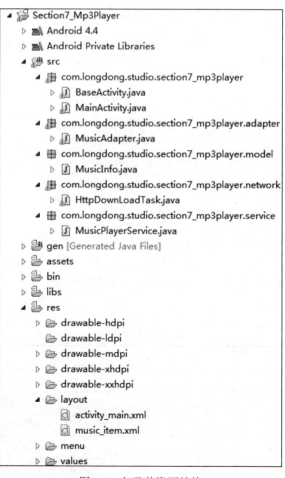

图 6-8 包及其资源结构

2. 命名空间

本示例设置了多个命名空间，分别用来保存用户界面、后台服务的源代码文件，具体说明见表 6-1。

表 6-1　命名空间

命名空间	说明
com.longdong.studio.section7_mp3player	存放与用户界面相关的源代码文件
com.longdong.studio.section7_mp3player.adapter	存放与页面适配器相关的源代码文件
com.longdong.studio.section7_mp3player.model	存放与数据管理相关的源代码文件
com.longdong.studio.section7_mp3player.network	存放与网络通信相关的源代码文件
com.longdong.studio.section7_mp3player.service	存放与服务相关的源代码文件

3. 源代码文件

源代码文件及说明见表 6-2。

表 6-2　源代码文件

包名称	文件名	说明
com.longdong.studio.section7_mp3player	BaseActivity.java	页面的基类
com.longdong.studio.section7_mp3player	MainActivity.java	主界面页的 Activity
com.longdong.studio.section7_mp3player.adapter	MusicAdapter.java	铃声列表的 Adapter
com.longdong.studio.section7_mp3player.model	MusicInfo.java	铃声数据管理
com.longdong.studio.section7_mp3player.network	HttpDownLoadTask.java	网络通信相关
com.longdong.studio.section7_mp3player.service	MusicPlayerService.java	音乐播放服务

4. 资源文件

Android 的资源文件保存在/res 的子目录中。

- /res/drawable 目录：保存的是图像文件。
- /res/layout 目录：保存的是布局文件。
- /res/values 目录：保存的是用来定义字符串和颜色的文件。

资源文件及说明见表 6-3。

表 6-3　资源文件

资源目录	文件	说明
drawable	back.jpg	主界面背景图片
drawable	down.png	下载按钮图标
drawable	ic_launcher.png	程序图标文件
drawable	music.png	播放区图标
drawable	pause.png	暂停按钮图标
drawable	play.png	播放按钮图标

续表

资源目录	文件	说明
layout	activity_main.xml	主界面的布局
	music_item.xml	列表行的布局
values	dimens.xml	保存尺寸的 XML 文件
	strings.xml	保存字符串的 XML 文件
	styles.xml	保存样式的 XML 文件

6.2.3 主要方法流程设计

音乐下载流程图如图 6-9 所示。

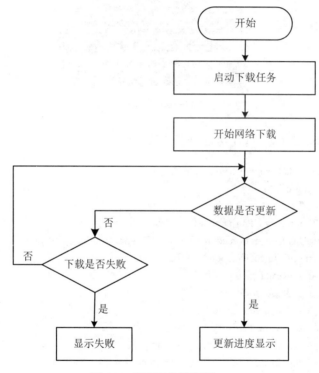

图 6-9　音乐下载流程图

6.3　代码实现

6.3.1　显示界面布局

系统主界面是进入系统后显示的界面，该界面包括一个 ListView、一个 ImageView、两个 TextView 和两个 Button，如图 6-10 所示。

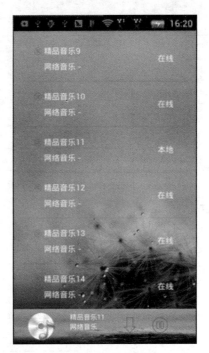

图 6-10 系统主界面

6.3.2 HttpURLConnection 网络通信方法实现

```
protected String doInBackground(String... params) {
        publishProgress(1);
        try {
            URL u = new URL(params[0]);
            HttpURLConnection conn = (HttpURLConnection) u.openConnection();
            conn.setConnectTimeout(10000);
            conn.setReadTimeout(10000);
            conn.setRequestMethod("GET");
            int len = conn.getContentLength();
            int code = conn.getResponseCode();
            InputStream in = conn.getInputStream();
            FileOutputStream out = new FileOutputStream(params[1]);
            byte[] buf = new byte[1024];
            int total = 0;
            float percent = 0;
            while (true) {
                int l = in.read(buf);

                if (l == -1) {
                    break;
                }
                total +=l;
                percent = (float)(total * 100) /(float) len;
```

```java
                publishProgress((int)percent);
                if(total>=len){
                    publishProgress((int)100);
                }
                out.write(buf, 0, l);
            }
            return params[1];
        } catch (MalformedURLException e) {
            //e.printStackTrace();
            if(taskResult!=null){
                taskResult.failed("URL地址不正确！");
            }
        } catch (IOException e) {
            //e.printStackTrace();
            if(taskResult!=null){
                taskResult.failed("网络不给力，请检查网络！");
            }
        }

        return null;
    }
```

6.3.3 XML 数据解析方法实现

下面的代码利用 DocumentBuilder 对 XML 文件进行解析，该方式属于 DOM。

```java
private List<MusicInfo> parseXML() {
    List<MusicInfo> list = new ArrayList<MusicInfo>();
    try {
        DocumentBuilderFactory factory = DocumentBuilderFactory.newInstance();
        DocumentBuilder builder = factory.newDocumentBuilder();
        //使用getAssets()方法从assets文件夹中读取文件流
        Document doc = builder.parse(getAssets().open("music.xml"));
        Element root = doc.getDocumentElement();
        NodeList nodeList = root.getElementsByTagName("Music");
        for (int i = 0; i < nodeList.getLength(); i++) {
            Node node = nodeList.item(i);

            if (node.getNodeType() == Node.ELEMENT_NODE) {
                MusicInfo info = new MusicInfo();
                Element e = (Element) node;
                String id = e.getAttribute("id");
                String type = e.getAttribute("type");
                String title = e.getAttribute("title");
                String singer = e.getAttribute("singer");
                String album = e.getAttribute("album");
                String url = e.getAttribute("url");
                String size = e.getAttribute("size");
```

```
                    info.setAlbum(album);
                    info.setId(id);
                    info.setType(type);
                    info.setSinger(singer);
                    info.setTitle(title);
                    info.setUrl(HOST+url);
                    info.setSize(Integer.valueOf(size));
                    list.add(info);
                }

            }
        } catch (FileNotFoundException e) {
            e.printStackTrace();
        } catch (IOException e) {
            e.printStackTrace();
        } catch (ParserConfigurationException e) {
            e.printStackTrace();
        } catch (SAXException e) {
            e.printStackTrace();
        }
        return list;
    }
```

6.4 关键知识点解析

6.4.1 AsyncTask（异步任务）的使用

我们知道 Android 的所有 UI 控件都运行在 UI 主线程中，因此当使用比较耗时的操作时（如网络请求、数据库读写、文件存储等），如果占用 UI 主线程，则用户操作将可能被阻塞，这不但严重影响用户体验，还可能被 Android 系统警告为程序暂无响应，要求用户等待或强制关闭。因此我们需要一种方式能够使上述耗时操作运行在非 UI 线程中，Android 本身为我们提供了 AsyncTask，它是一个抽象类 public abstract class AsyncTask <Params, Progress, Result>，我们需要继承 AsyncTask 以实现自定义的异步任务，比如：

public class LoadImageTask extends AsyncTask<String,Integer, Bitmap>

继承 AsyncTask 时还需要定义其中的三个泛型<Params, Progress, Result>。
三种泛型类型说明如下：
- Params：主线程执行 Task 时的传入参数，将传递给后台线程。
- Progress：后台线程执行的进度。
- Result：后台线程计算结果的类型。

某些时候并不是三个参数都被使用，如果不需要使用，可以用 java.lang.Void 类型代替。
一个 AsyncTask 的执行一般包括以下几个步骤：
（1）execute(Params... params)：开始执行一个异步任务（起点）。

（2）onPreExecute()：在 execute 被调用后立即执行，一般用来在执行后台任务前在 UI 中做一些提示。

（3）doInBackground(Params... params)：在 onPreExecute()完成后立即执行，该方法运行在独立的工作线程中，因此用于执行耗时操作。此方法将接收 execute 调用时的传入参数（params），并返回计算结果。在执行过程中可以调用 publishProgress 来传递进度信息。

（4）onProgressUpdate(Progress... values)：接收由 publishProgress(Progress... values)从后台线程传递的进度信息，在该方法中可直接操作 UI，将进度信息更新到 UI 组件上。

（5）onPostExecute(Result result)：当后台操作结束时，此方法将会被调用，计算结果将作为参数传递到此方法中。

具体代码实现可参照如下代码：

```java
public class LoadImageTaskModel extends AsyncTask<String,Integer,Bitmap> {
    private Activity mContext;
    public LoadImageTaskModel(Activity context){
        mContext = context;
    }
    protected void onPreExecute(){
        //UI界面可以在此刷新
    }
    protected Bitmap doInBackground(String...params) {
        //在后台线程中加载Bitmap
        Bitmap bitmap = Bitmap.createBitmap(0,0,null);
        return bitmap;
    }
    protected void onPostExecute(Bitmap result) {
        //后台进程将处理结果传递到主线程
        //此处可以刷新UI了
    }
    protected void onProgressUpdate(Integer... values) {
        //刷新UI，更新进度
    }
}
```

6.4.2 HttpClient、HttpURLConnection、okHttp 和 Volley 的网络通信对比

HttpClient、HttpURLConnection、okHttp 和 Volley 这四种网络通信方案是伴随着 Android 平台的成长而不断演进的，HttpClient 和 HttpURLConnection 作为支撑 Android HTTP 网络通信的元老，在 Android 发布初期就扮演着举足轻重的角色，HttpClient 作为 Apache 基金会的开源项目及在 Java 业界知名度极高的 HTTP 通信工具，拥有强大的 API 支撑和扩展 jar 包，也拥有出色的稳定性，在 Android 4.0 之前是开发者不二的选择。而随着谷歌对 HttpURLConnection 的不断优化和 bug 的修正，HttpURLConnection 逐渐成为 Android 4.0 后的较佳选择，尽管其 API 不如 HttpClient 丰富，但由于其轻量的特点，开发者也更容易对其进行功能扩展。HttpURLConnection 获得了谷歌官方的推荐，而 HttpClient 逐步被谷歌弃用，在 Android 5.0 中 HttpClient 的主要类和方法被标示为废弃，在 Android 6.0 中已经删除了

HttpClient 的相关包。不过不必过于担心，如果开发者希望在 Android 6.0 中继续使用 HttpClient，可以采用下列方法：如果使用 Eclipse，需要在 libs 中加入 org.apache.http.legacy.jar，这个 jar 包在 Android 6.0 的 SDK 下载后，可以在**\android-sdk-windows\platforms\android-23\optional 目录下找到。如果使用 Android Studio，需要在相应的 module 下的 build.gradle 中加入如下代码：

```
android {
    useLibrary 'org.apache.http.legacy'
}
```

当 Android 演进至 4.4 版本时，okHttp 正式被谷歌吸收进来，谷歌用 okHttp 对 HttpURLConnection 的底层进行了重新实现。okHttp 是一个高效的 HTTP 客户端，其支持 SPDY 协议，可以合并多个到同一个主机的请求，使用连接池技术减少请求的延迟，开启 gzip 压缩减少传输的数据量，采用缓存响应避免重复的网络请求。尤其在网络出现繁忙或信号不稳定的时候，okHttp 可以比传统的 HttpClient 和 HttpURLConnection 处理得更好，对于请求失败及 https 握手失败等问题，okHttp 亦拥有专门的处理方案。因此在当前的 Android 开发中，开发者既可以直接使用由 okHttp 作为底层实现 HttpURLConnection，又可以引入 com.squareup.okhttp:okhttp 和 com.squareup.okio:okio 两个 jar，直接使用 okHttp 的相应 API。

由于 Android 中进行网络通信需要借助线程和 Handler 的消息机制，所以开发者无论使用 HttpClient、HttpURLConnection 还是 okHttp，都需要在 UI 线程中获取请求数据内容，在工作线程中构建网络请求并接收响应，最后将响应的数据结果回传给 UI 线程进行显示，这样就增加了多个开发环节，增加了开发难度，如果开发者封装调度得不好，还可能产生严重的性能问题。因此，谷歌为开发者提供了 Volley 这个简化版的网络任务库，其负责处理请求、响应、线程、缓存、回调 UI 等问题。它不但可以直接处理 JSON、文本，还可以缓存图片，同时也支持一定程度的自定义扩展。使用 Volley 可以替代 Handler 或 AsyncTask，进而使程序更加简便，编码更加轻松。Volley 在 Android 2.3 及以上版本默认利用 HttpURLConnection 实现 HTTP 网络通信，而在 Android 2.2 及以下版本默认使用 HttpClient 实现 HTTP 网络通信。

注意：SPDY 是谷歌开发的基于 TCP 的应用层协议，用以最小化网络延迟，提升网络速度，优化用户的网络使用体验。SPDY 并不是一种用于替代 HTTP 的协议，而是对 HTTP 的增强。新协议的功能包括数据流的多路复用、请求优先级以及 HTTP 报头压缩。谷歌表示，引入 SPDY 协议后，在实验室测试中页面加载速度比原先快 64%。

6.4.3　HttpClient 和 HttpURLConnection 的使用方法

1. HttpClient 的使用方法

HttpClient 的使用方法可以分为如下步骤：

（1）创建 HttpClient 实例。

HttpClient client = new DefaultHttpClient();

（2）确定请求方法（Get or Post）类型。

HttpGet request = new HttpGet(uri);

HttpPost request = new HttpPost(uri);

（3）如果是 Post 类型，设置请求实体。

request.setEntity(Entity);

（4）执行 execute 方法，获取服务器 HTTP 响应。
HttpResponse response = client.execute(request);

（5）获取响应的相关信息，如响应码、响应首部、响应主体等。
long contentLength = response.getEntity().getContentLength();
Header contentType = response.getEntity().getContentType();
InputStream in = response.getEntity().getContent();

（6）读取首部响应主体的输入流，转换成所需要的数据形式。

具体代码如下：

```
try{
    HttpClient client = new DefaultHttpClient();
    HttpGet request = new HttpGet(uri);
    HttpResponse response = client.execute(request);
    long contentLength = response.getEntity().getContentLength();
    Header contentType = response.getEntity().getContentType();
    in = response.getEntity().getContent();
    out = new ByteArrayOutputStream();
    byte[] buffer = new byte[1024];
    while(true){
        int len = in.read(buffer);
        if(len == -1){break;}
        out.write(buffer,0.len);
    }
    result = out.toByteArray();
} catch (ClientProtocolException e) {}
```

2. HttpURLConnection 的使用方法

HttpURLConnection 的使用方法可以分为如下步骤：

（1）创建 HttpURLConnection 实例。
URL url = new URL(address);
HttpURLConnection conn = (HttpURLConnection)url .openConnection();

（2）设置超时时间、首部、请求方法。
conn.setConnectTimeout(10000);
conn.setRequestProperty("Accept", "image/jpeg,*/*");
conn.setRequestProperty("Connection", "close");
conn.setRequestMethod("GET");

（3）获取服务器响应的输入流。
InputStream in = conn.getInputStream();

（4）向服务器写入请求体或者上传数据。
OutputStream out = conn.getOutputStream();

具体代码如下：

```
HttpURLConnection conn = (HttpURLConnection)source.openConnection();
conn.setConnectTimeout(10000);      //与服务器创建连接的超时时间
conn.setReadTimeout(10000);
conn.setRequestMethod("GET");       //等待服务器响应的时间
```

```
conn.setRequestProperty("Accept","image/jpeg,*/*");
conn.setRequestProperty("Connection","close");
long size = conn.getContentLength();
InputStream in = conn.getInputStream();
//OutputStream outToServer = conn.getOutputStream();
ByteArrayOutputStream out = new ByteArrayOutputStream();
byte[] buffer = new byte[1024];
    while(true){
        int len = in.read(buffer);
        if(len == -1){break;}
        out.write(buffer,0,len);
    }
in.close();
out.close();
return out.toByteArray();
```

6.5 问题与讨论

1. 利用 HttpClient 通过 Post 请求提交数据时，如何改变默认编码？
2. Get 与 Post 请求的区别是什么？

项目 7 新闻客户端

本项目在简易网络音乐播放器项目的基础上,对网络交互和数据解析进行进一步学习。
项目需求描述如下:
(1) 实现一个新闻客户端。
(2) 从服务器获取数据并显示。
(3) 实现新闻的列表。
(4) 实现详细页面。
(5) 用户体验设计 RSS 服务。

(1) 使用 DOM 方式解析 XML 数据。
(2) 利用 WebView 显示 HTML 页面。
(3) 深入理解 XML 数据格式。
(4) 利用 ViewHolder 优化 AdapterView。
(5) 掌握 Fragment 的简单使用方法(FragmentStatePagerAdapter)。
(6) 菜单的使用技巧(ActionBar)。

7.1 总体设计

新闻客户端应实现以下功能:新闻列表显示、新闻详细内容查看。
整个程序除总体模块外,主要分为基础架构模块、用户界面模块、数据管理与控制模块和网络通信模块四大部分。在整个系统中,基础架构模块提供各项基础功能供其他模块调用,总体模块控制系统的生命周期,用户界面模块负责界面显示,数据管理与控制模块主要提供数据管理功能,网络通信模块负责与服务器通信。

7.1.1 功能模块框图

根据总体分析结果可以总结出功能模块框图,如图 7-1 所示。
总体模块的作用主要是生成应用程序的主类,控制应用程序的生命周期;基础架构模块主要提供程序架构、所有 Activity 公用的父类、所有 Activity 公用的方法,包括自定义风格对话框、自定义提示框等功能；数据管理与控制模块主要提供数据获取、数据解析、数据组织和数据缓存功能；用户界面模块包括新闻列表、新闻详情等功能；网络通信模块主要负责从服务端获取新闻数据信息，数据从指定网络地址的服务器通过 HTTP 方式下载获取。

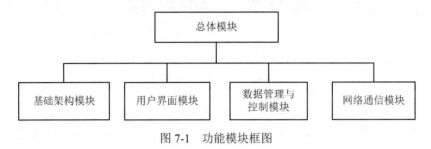

图 7-1　功能模块框图

7.1.2　系统流程图

根据总体分析结果及功能模块框图梳理出系统启动的主要流程，如图 7-2 所示。

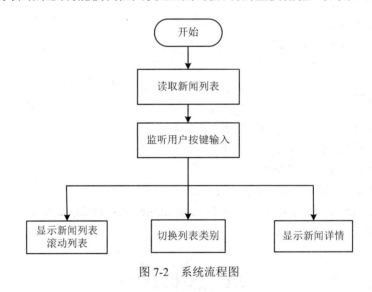

图 7-2　系统流程图

7.1.3　界面设计

本应用是显示新闻的应用，主要实现新闻列表显示和新闻详细内容显示。

根据程序功能需求，可以规划出软件的主要界面如下：

（1）新闻列表：启动应用程序后显示新闻列表。

（2）新闻详细内容：用户在列表中选中新闻后进入详细页，显示新闻的详细内容。

系统主界面如图 7-3 所示。

从图 7-3 中可以很直观地看到，列表页界面划分了两个区域：新闻栏目区和新闻列表区。详细页使用 WebView 显示新闻内容，底部有放大/缩小按钮和看法文本输入框。

图 7-3　系统主界面

7.2　详细设计

7.2.1　模块描述

在系统总体分析及界面布局设计完成后，主要工作就转入对各个功能模块的详细设计阶段。

1. 基础架构模块详细设计

基础架构模块主要提供程序架构、所有 Activity 公用的父类、所有 Activity 公用的方法，包括自定义风格对话框、自定义提示框等功能。

基础架构模块功能如图 7-4 所示。

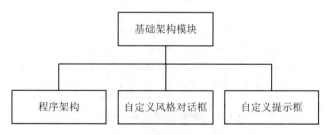

图 7-4　基础架构模块功能图

2. 用户界面模块详细设计

用户界面模块的主要任务是显示新闻列表、新闻详细内容、功能按钮、实现与用户的交互，即当用户点击按键或者屏幕的时候，监听器会去调用相应的处理办法或其他相应的处理模块。

本模块包括新闻列表显示、新闻详细信息显示、操作按钮等功能。

用户界面模块功能如图 7-5 所示。

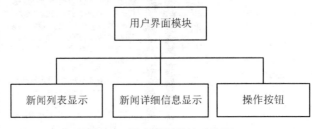

图 7-5　用户界面模块功能图

3. 数据管理与控制模块详细设计

数据管理与控制模块主要提供数据获取、数据解析、数据组织和数据缓存功能。

数据管理与控制模块和用户界面模块可以调用基础架构模块的一些通用方法。数据管理与控制模块为用户界面模块提供数据，同时可以接收并保存用户界面模块产生的数据。

数据管理与控制模块功能如图 7-6 所示。

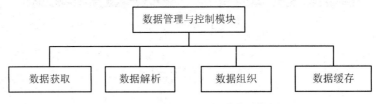

图 7-6　数据管理与控制模块功能图

4. 网络通信模块详细设计

网络通信模块根据用户界面的需求调用访问服务器，接收服务器返回的数据并解析，同时显示到用户界面。

本模块包括发送网络请求、接收网络应答、网络数据解析等功能。

网络通信模块功能如图 7-7 所示。

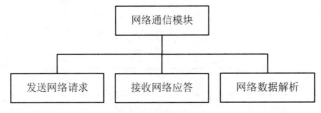

图 7-7　网络通信模块功能图

7.2.2　系统包及其资源规划

1. 文件结构

根据系统功能设计，本系统封装一个基础的 Activity 类，加载各个子页的通用控件，并提供一些基础的实现方法，例如设置进度条、标题等常用的方法，程序中的 Activity 都可继承此基类，继承后就可直接使用基类中封装的基础方法。

系统使用一个 Activity，用于显示新闻列表和栏目切换滚动条。包及其资源结构如图 7-8 所示。

图 7-8　包及其资源结构

2．命名空间

本示例设置了多个命名空间，分别用来保存用户界面、后台服务的源代码文件，具体说明见表 7-1。

表 7-1　命名空间

命名空间	说明
com.example.news.model	存放与新闻数据相关的源代码文件
com.example.news.utils	存放与新闻栏目相关的源代码文件
com.longdong.studio.section9.news	存放与用户界面相关的源代码文件

3．源代码文件

源代码文件及说明见表 7-2。

表 7-2 源代码文件

包名称	文件名	说明
com.example.news.model	NewsModel.java	新闻数据结构相关
com.example.news.network	HttpTask.java	网络异步通信相关
	HttpUtils.java	网络数据请求应答相关
Com.example.news.utils	Constants.java	常量数据类
com.longdong.studio.section9.news	BaseActivity.java	页面的基类
	MainActivity.java	主界面页的 Activity
	NewsListFragment.java	新闻列表页的 Activity
	SplashActivity.java	欢迎页的 Activity
	WebBrowserActivity.java	详细页的 Activity

4. 资源文件

Android 的资源文件保存在/res 的子目录中。
- /res/drawable 目录：保存的是图像文件。
- /res/layout 目录：保存的是布局文件。
- /res/values 目录：保存的是用来定义字符串和颜色的文件。

资源文件及说明见表 7-3。

表 7-3 资源文件

资源目录	文件	说明
drawable	ic_launcher.png	程序图标文件
	splash.jpg	启动页图片
layout	activity_main.xml	主界面的布局
	activity_splash.xml	启动页的布局
	chat_collection.xml	栏目的布局
	item_news.xml	新闻条目的布局
	news_list_fragment.xml	新闻列表碎片的布局
values	colors.xml	保存颜色的 XML 文件
	dimens.xml	保存尺寸的 XML 文件
	strings.xml	保存字符串的 XML 文件
	styles.xml	保存样式的 XML 文件

7.2.3 主要方法流程设计

查看新闻流程图如图 7-9 所示。

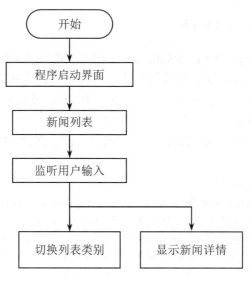

图 7-9　查看新闻流程图

7.3　代码实现

7.3.1　显示界面布局

1．系统主界面

系统主界面是进入系统后显示的新闻列表界面，该界面包括一个 PagerTabStrip 控件和一个 Fragment。Fragment 中嵌入了一个 ListView，每个 ListView 条目中包含一个 ImageView 和两个 TextView。主界面如图 7-10 所示。

2．新闻详细界面

新闻详细界面用于显示新闻详细信息，该界面包括一个 WebView，如图 7-11 所示。

图 7-10　系统主界面

图 7-11　新闻详细界面

7.3.2 RSS 内容读取方法实现

真正简易聚合（Really Simple Syndication，RSS）是基于可扩展标识语言（XML）文本的格式，它提供了一种更为方便、高效的互联网信息的发布和共享方法，用更少的时间分享更多的信息，目前广泛用于网上新闻频道。通常 RSS 文件都是标为 XML，RSS files（也被称为 RSS feeds 或者 channels）通常只包含简单的项目列表。一般而言，每一个项目都含有标题、摘要和 URL 链接，还有其他可选的信息，例如日期、创建者的名字等。

1. 创建通信任务

创建通信任务的代码如下：

```java
public View onCreateView(LayoutInflater inflater, ViewGroup container, Bundle savedInstanceState) {
    final MainActivity parentActivity = (MainActivity ) getActivity();
    View rootView = inflater.inflate(R.layout.news_list_fragment, null);
    newsListView = (ListView) rootView.findViewById(R.id.newsList);

    if (mNewsList == null) {
        mNewsList = new LinkedList<NewsModel>();
    }
    final NewsListAdapter adapter = new NewsListAdapter(getActivity(), mNewsList);
    newsListView.setAdapter(adapter);
    newsListView.setOnItemClickListener(adapter);

    RequestData data = new RequestData();
    data.uri = Constants.RSS_ADDRESS.get(column);
    parentActivity.startHttpTask(new TaskResultListener() {
        @Override
        public void result(ResposneBundle b) {
            Log.e("result",b.getContent());
            mNewsList.clear();
            mNewsList.addAll(parseXML(b.getContent()));
            adapter.notifyDataSetChanged();
        }
        @Override
        public void failed(final String message) {
            newsListView.post(new Runnable() {
                @Override
                public void run() {
                    parentActivity.showAlertDialog("出错了",message);
                }
            });
        }
    }, data);
    return rootView;
}
```

2. 显示 splash 页

显示 splash 页的代码如下：

```java
public class SplashActivity extends Activity {
    @Override
    protected void onCreate(Bundle savedInstanceState) {
        super.onCreate(savedInstanceState);
        requestWindowFeature(android.view.Window.FEATURE_NO_TITLE);
        setRequestedOrientation(ActivityInfo.SCREEN_ORIENTATION_PORTRAIT);
        setContentView(R.layout.activity_splash);
        View view = findViewById(R.id.title);
        view.postDelayed(new Runnable() {

            @Override
            public void run() {
                Intent in = new Intent(SplashActivity.this,MainActivity.class);
                startActivity(in);
                overridePendingTransition(android.R.anim.slide_in_left,
                    android.R.anim.slide_out_right);
                finish();
            }
        }, 1500);
    }
}
```

3. RSS 数据地址

下面的代码中保存了 RSS 数据地址。

```java
public class Constants {
    //国内要闻
    public final static String CHINA_FOCUS = "http://rss.sina.com.cn/news/china/focus15.xml";
    //国际要闻
    public final static String WORLD_FOCUS = "http://rss.sina.com.cn/news/world/focus15.xml";
    //科技要闻
    public final static String TECH_FOCUS = "http://rss.sina.com.cn/tech/rollnews.xml";
    //社会万象
    public final static String SOCIETY = "http://rss.sina.com.cn/news/society/misc15.xml";
    //体育要闻
    public final static String SPORTS = "http://rss.sina.com.cn/roll/sports/hot_roll.xml";
    //奇闻轶事
    public final static String WONDER = "http://rss.sina.com.cn/news/society/wonder15.xml";
    //精彩图片
    public final static String PHOTO = "http://rss.sina.com.cn/sports/global/photo.xml";
    public static HashMap<String,String> RSS_ADDRESS = new HashMap<String,String>(){
        {
            put("国内要闻", CHINA_FOCUS);
            put("国际要闻", WORLD_FOCUS);
            put("科技要闻", TECH_FOCUS);
            put("社会万象", SOCIETY);
```

```
            put("体育要闻", SPORTS);
            put("奇闻轶事", WONDER);
            put("精彩图片", PHOTO);
        }
    };
    public static String[] COLUMN_ORDER = new String[]{"国内要闻","国际要闻","科技要闻","社会万象",
"体育要闻","奇闻轶事","精彩图片"};
}
```

4．parseXML 函数

parseXML 函数主要用于解析从 RSS 返回的 XML 文件，代码如下：

```
private List<NewsModel> parseXML(String content){
        List<NewsModel> list = new ArrayList<NewsModel>();
        DocumentBuilderFactory factory = DocumentBuilderFactory.newInstance();
        try {
            DocumentBuilder builder = factory.newDocumentBuilder();
            ByteArrayInputStream in = new ByteArrayInputStream(content.getBytes());
            Document doc = builder.parse(in);
            Element root = doc.getDocumentElement();

            Element channelNode = (Element) root.getElementsByTagName("channel").item(0);

            if(channelNode == null){
                return list;
            }
            NodeList itemNodeList = channelNode.getElementsByTagName("item");
            for(int i =0 ; i<itemNodeList.getLength();i++){
                Node node = itemNodeList.item(i);
                if(node.getNodeType() ==Node.ELEMENT_NODE){
                    NewsModel model =new NewsModel();
                    Element itemElement = (Element)node;
                    Element authorElement = (Element)
                            itemElement.getElementsByTagName("author").item(0);
                    Element commentsElement = (Element)
                            itemElement.getElementsByTagName("comments").item(0);
                    Element descriptionElement = (Element)
                            itemElement.getElementsByTagName("description").item(0);
                    Element pubDateElement = (Element)
                            itemElement.getElementsByTagName("pubDate").item(0);
                    Element titleElement = (Element)
                            itemElement.getElementsByTagName("title").item(0);
                    Element linkElement = (Element)
                            itemElement.getElementsByTagName("link").item(0);
                    model.setAuthor(authorElement.getTextContent());
                    model.setComments(commentsElement.getTextContent());
                    model.setDescription(descriptionElement.getTextContent());
                    model.setPubDate(pubDateElement.getTextContent());
```

```
                model.setTitle(titleElement.getTextContent());
                model.setLink(linkElement.getTextContent());
                list.add(model);
            }
        }
    } catch (ParserConfigurationException e) {
        e.printStackTrace();
    } catch (IOException e) {
        e.printStackTrace();
    } catch (SAXException e) {
        e.printStackTrace();
    }
    return list;
}
```

7.3.3 利用 WebView 显示 HTML 页面

WebView 可以使网页轻松地内嵌到 App 里，代码如下：

```
public void onItemClick(AdapterView<?> adapterView, View v, int position, long itemId) {
    NewsModel n = this.getItem(position);
    Intent in = new Intent(mContext,WebBrowserActivity.class);
    in.putExtra("title", n.getTitle());
    in.putExtra("url", n.getLink());
    mContext.startActivity(in);
    ((Activity) mContext).overridePendingTransition(android.R.anim.slide_in_left,
            android.R.anim.slide_out_right);
}
```

7.3.4 利用 ViewHolder 优化 AdapterView

如果对效率要求比较高的话可以采用这种办法，唯一的缺点是多了一个内部类 ViewHolder，代码如下：

```
@Override
public View getView(int position, View convertView, ViewGroup parent) {
    ViewHolder holder = null;
    if (convertView == null) {
        convertView = LayoutInflater.from(mContext).inflate(R.layout.item_news, null);
        holder = new ViewHolder();
        holder.title = (TextView) convertView.findViewById(R.id.title);
        holder.des = (TextView) convertView.findViewById(R.id.des);
        holder.icon = (ImageView) convertView.findViewById(R.id.icon);
        convertView.setTag(holder);
    } else {
        holder = (ViewHolder) convertView.getTag();
    }
    //使用随机数为图片指定随机背景颜色
```

```java
            int avatarColor =
                Color.argb(255, random.nextInt(256), random.nextInt(256), random.nextInt(256));
            holder.icon.setBackgroundColor(avatarColor);
            holder.title.setText(this.getItem(position).getTitle());
            holder.des.setText(this.getItem(position).getDescription());
            return convertView;
}
```

7.3.5　Fragment 的简单使用方法（FragmentStatePagerAdapter）

在项目 2 中我们曾经学习过 ViewPager 的使用方法，而谷歌设计 ViewPager 的初衷更多地在于配合 Fragment 进行使用。

```java
@Override
protected void onConentViewLoad(ViewGroup container) {
        this.setContentView(R.layout.chat_collection);
        mPager = (ViewPager) findViewById(R.id.pager);
        mPager.setAdapter(new NewsPageFragmentAdapter(getSupportFragmentManager()));
        mPager.setPageMargin(16);

        pagerTabStrip = (PagerTabStrip) findViewById(R.id.pager_tab_strip);
        pagerTabStrip.setTabIndicatorColorResource(R.color.tab_indicator_color);
        pagerTabStrip.setTextColor(getResources().getColor(R.color.tab_text_color));
}
@Override
public void onBackPressed() {
super.onBackPressed();
}

public class NewsPageFragmentAdapter extends FragmentStatePagerAdapter {
    public NewsPageFragmentAdapter(FragmentManager fm) {
        super(fm);
    }

    @Override
    public Fragment getItem(int position) {
      Fragment frag = new NewsListFragment(Constants.COLUMN_ORDER[position]);
      return frag;
    }

    @Override
    public CharSequence getPageTitle(int position) {
    return Constants.COLUMN_ORDER[position] ;
    }

    @Override
    public int getCount() {
      return Constants.COLUMN_ORDER.length;
```

            }
        }

7.3.6 菜单的使用技巧（ActionBar）

```
@Override
public boolean onCreateOptionsMenu(Menu menu) {
    int idx = 0;
     for(String title:Constants.COLUMN_ORDER){
         menu.add(Menu.NONE, idx, idx, title);
         idx++;
     }
     return true;
}

@Override
public boolean onOptionsItemSelected(MenuItem item) {
    mPager.setCurrentItem(item.getItemId(), true);
    return super.onOptionsItemSelected(item);
}
```

7.4 关键知识点解析

7.4.1 用户体验

1. WebView 页面意外被杀

为了防止 WebView 意外被杀死，可以在 onSaveInstanceState(Bundle)中调用 webview.saveState(bundle)保存状态，然后在 onCreate(Bundle saveInstanceState)里通过 savedInstanceState == null 判断，如果不为 null，即可通过 webview.restoreState(bundle)恢复。

2. WebView 返回时跳转到上次的阅读进度

正常的返回不会触发 onSaveInstanceState()方法。如果需要返回时跳转到上次的阅读进度，需要通过 webview.getContentHeight()和 webview.getSrollY()得到滚动位置所占 HTML 页面实际内容长度的比例。由于 HTML 加载内容可能显示不完全，getContentHeight()的值很有可能是会变化的，所以我们最好算出这个百分比，以便下次恢复的时候也能根据百分比再滚动，关键代码如下：

```
private float scrollPercentage(WebView webview) {
    float webviewTopView = webview.getTop();
    float webviewHeight = webview.getContentHeight();
    float webviewScrollPosition = webview.getScrollY();
    float percentWebview = (webviewScrollPosition - webviewTopView ) / webviewHeight ;
    return percentWebview;
}
```

然后在 onDestory()里保留如下值：
@Override

```
protected void onDestroy() {
    mScrollPercentage= scrollPercentage(webview);
    super.onDestroy();
}
```

最后在 WebViewClient.onPageFinished()中跳转到 mScrollPercentage 进度，代码如下：

```
@Override
public void onPageFinished(WebView view, String url) {

    if (mCrrentScrollPercentage&& 0<mScrollPercentage) {
            mCrrentScrollPercentage= false;
            view.postDelayed(new Runnable() {
                @Override
                public void run() {
                    float webviewsize = webview.getContentHeight() - webview.getTop();
                    float positionInWebView = webviewsize * mScrollPercentage;
                    int positionY = webview.getTop() + positionInWebView ;
                    webview.scrollTo(0, positionY);
                }
            }, 200);
    }

    super.onPageFinished(view, url);
}
```

7.4.2 RSS 阅读器实现

1. 设计思路

（1）UI。下拉列表选择 RSS 的源 ListView 显示 RSS 源中读取的新闻列表。

（2）相关功能。

1) 支持手动更新，选择下拉列表 RSS 源，ListView 内容自动更新。

2) 点击 ListView 项，显示对应的新闻详情。

（3）功能菜单。

1) 退出：用来退出整个应用程序。

2) 更新新闻：加载最新 RSS 新闻。

2. 实现过程

（1）RSS 解析。根据 7.3.2 节中的步骤实现 RSS 解析。

（2）界面实现。

主界面布局文件 main_activity.xml 代码如下：

```xml
<?xml version="1.0" encoding="utf-8"?>
<android.support.constraint.ConstraintLayout xmlns:android="http://schemas.android.com/apk/res/android"
    xmlns:app="http://schemas.android.com/apk/res-auto"
    xmlns:tools="http://schemas.android.com/tools"
    android:layout_width="match_parent"
    android:layout_height="match_parent"
```

```
        app:layout_behavior="@string/appbar_scrolling_view_behavior"
        tools:context=".MainActivity"
        tools:showIn="@layout/activity_main">


    <TextView
        android:id="@+id/textView"
        android:layout_width="wrap_content"
        android:layout_height="wrap_content"
        android:text="选择新闻"
        tools:layout_editor_absoluteX="0dp"
        tools:layout_editor_absoluteY="2dp" />

    <Spinner
        android:id="@+id/spinner"
        android:layout_width="match_parent"
        android:layout_height="22dp"
        android:layout_marginStart="8dp"
        app:layout_constraintStart_toEndOf="@+id/textView"
        tools:layout_editor_absoluteY="2dp" />

    <ListView
        android:layout_width="match_parent"
        android:layout_height="395dp"
        android:layout_marginEnd="8dp"
        android:layout_marginStart="8dp"
        android:layout_marginTop="8dp"
        app:layout_constraintEnd_toEndOf="parent"
        app:layout_constraintStart_toStartOf="parent"
        app:layout_constraintTop_toBottomOf="@+id/spinner" />
```

`</android.support.constraint.ConstraintLayout>`

List 数据项布局文件 list_item.xml 代码如下：

```
<?xml version="1.0" encoding="utf-8"?>
<TextView xmlns:android="http://schemas.android.com/apk/res/android"
    android:id="@+id/title_name" android:layout_width="wrap_content"
    android:layout_height="wrap_content"/>
```

菜单布局文件 menu.xml 代码如下：

```
<?xml version="1.0" encoding="UTF-8"?>
<menu xmlns:android="http://schemas.android.com/apk/res/android">
    <item android:id="@+id/refresh" android:title="更新新闻"></item>
    <item android:id="@+id/exit" android:title="退出"></item>
</menu>
```

（3）主要代码实现。

具体代码如下：

```
package com.kfxmsxjc.rss;
```

```java
public class RssMainActivity extends ListActivity {

    private Spinner rssSourceSpinner;
    private String[] rssUrlArray;
    private String mRssUrl;
    private List<Message> messageList;

    @Override
    public void onCreate(Bundle savedInstanceState) {
        super.onCreate(savedInstanceState);
        setContentView(R.layout.main);
        rssSourceSpinner = (Spinner) findViewById(R.id.rssSourceSpinner);

        rssUrlArray = getResources().getStringArray(R.array.source_url);

        ArrayAdapter<CharSequence> spinnerAdapter = ArrayAdapter.createFromResource(this,
            R.array.rss_title, R.layout. list_item);
        spinnerAdapter.setDropDownViewResource(android.R.layout.simple_spinner_dropdown_item);
        rssSourceSpinner.setAdapter(spinnerAdapter);
        ssSelectSpinner.setOnItemSelectedListener(new OnItemSelectedListener() {
            public void onItemSelected(AdapterView<?> parent, View view, int pos, long id) {
                mRssUrl = rssUrlArray[pos];
            }

            public void onNothingSelected(AdapterView parent) {
            }
        });
        ssSelectSpinner.set
    }

    @Override
    public boolean onCreateOptionsMenu(Menu menu) {
        MenuInflater inflater = getMenuInflater();
        inflater.inflate(R.menu.menu, menu);
        return true;
    }

    @Override
    public boolean onOptionsItemSelected(MenuItem item) {
        int item_id = item.getItemId();
        switch (item_id) {
            //更新新闻
            case R.id.refresh:
                loadRss();
                break;
            //退出
```

```
            case R.id.exit:
                finish();
                break;
        }
        return true;
    }

    /*
     * 根据现在选中的spinner来加载对应的新闻
     */
    public void loadRss() {

            RssParser rssparser = new RssParser (mRssUrl);
            messageList = rssparser.parse();

            List<String> titles = new ArrayList<String>(messageList.size());
            for (Message msg : this.messageList) {
                titles.add(msg.getTitle());
            }
            ArrayAdapter<String> adapter = new ArrayAdapter<String>(this,
                    R.layout. list_item , titles);
            this.setListAdapter(adapter);

    }

    @Override
    protected void onListItemClick(ListView l, View v, int position, long id) {
        super.onListItemClick(l, v, position, id);
        Intent intent= new Intent(Intent.ACTION_VIEW,
                Uri.parse(this.messageList.get(position).getUrl()));
        startActivity(intent );
    }
}
```

7.4.3 深入理解 XML 数据格式

可扩展标记语言（Extensible Markup Language，XML）是一种标记语言，很类似 HTML，它的设计宗旨是传输数据，而非显示数据。XML 标签没有被预定义，需要自行定义标签。XML 被设计为具有自我描述性，是 W3C 的推荐标准。XML 仅仅是纯文本，有能力处理纯文本的软件都可以处理 XML。

XML 和 HTML 为不同的目的而设计。XML 被设计用来传输和存储数据，其焦点是数据的内容。HTML 被设计用来显示数据，其焦点是数据的外观。HTML 旨在显示信息，而 XML 旨在传输信息。

没有任何行为的 XML 是不作为的。也许这有些难以理解，但是 XML 不会做任何事情。XML 被设计用来结构化、存储以及传输信息。

新闻存储为 XML，代码如下：
```
<news>
<title>标题</title>
<Summary>摘要</Summary>
<url>www.xxx.com</url>
</news>
```
上面的这条新闻有自我描述性。它拥有标题，同时包含了具体详情链接和摘要信息。但是，这个 XML 文档仍然没有做任何事情。它仅仅是包装在 XML 标签中的纯粹信息。我们需要编写软件或者程序，才能传送、接收和显示出这个文档。XML 的语法规则很简单，且很有逻辑。

在 XML 中，省略关闭标签是非法的，所有元素都必须有关闭标签，格式如下：
```
//错误格式
    <p>王
    <p>李


//正确格式
    <p>王</p>
    <p>李</p>
```
XML 标签区分大小写，在 XML 中，标签<Letter>与标签<letter>是不同的。

注意：为了保证 XML 不会出现乱码，需要保存时的编码和设置打开时的编码一致。

XML 的三种解析方式：

（1）DOM 解析器把 XML 转换为 JavaScript 可存取的对象，需要一次性加载 XML 文件，再使用 DOM 的 API 去进行解析，这样会很大程度地消耗内存，对性能会有一定影响。所以在 Android 开发中，不太推荐使用 DOM 的方式来解析和操作 XML。DOM 解析示例如下：

```java
package com.jpy.myapplication;

import java.io.InputStream;
import java.util.ArrayList;
import java.util.List;
import javax.xml.parsers.DocumentBuilder;
import javax.xml.parsers.DocumentBuilderFactory;
import org.w3c.dom.Document;
import org.w3c.dom.Element;
import org.w3c.dom.Node;
import org.w3c.dom.NodeList;


public class DomboyService {

    public List getboys(InputStream stream) throws Throwable {
        List list = new ArrayList();
        DocumentBuilderFactory factory = DocumentBuilderFactory.newInstance();
        DocumentBuilder builder = factory.newDocumentBuilder();
        Document dom = builder.parse(stream);
```

```java
//解析完成，并以DOM树的方式存放在内存中，比较消耗性能
//开始使用DOM的API去解析
Element root = dom.getDocumentElement();
//根元素
NodeList boyNodes = root.getElementsByTagName("boy");
//返回所有的boy元素节点
//开始遍历
for (int i = 0; i < boyNodes.getLength(); i++) {
    Boy boy = new Boy();
    Element boyElement = (Element) boyNodes.item(i);
    boy.setId(boyElement.getAttribute("id"));
    //将boy元素节点的属性节点ID的值赋给boy对象
    NodeList boyChildrenNodes = boyElement.getChildNodes();
    //获取boy元素节点的所有子节点
    //遍历所有子节点
    for (int j = 0; j < boyChildrenNodes.getLength(); j++) {
        //判断子节点是否是元素节点（如果是文本节点，可能是空白文本，不处理）
        if (boyChildrenNodes.item(j).getNodeType() == Node.ELEMENT_NODE) {
            //子节点——元素节点
            Element childNode = (Element) boyChildrenNodes.item(j);
            if ("name".equals(childNode.getNodeName())) {
                //如果子节点的名称是name，将子元素节点的第一个子节点的值赋给boy对象
                boy.setName(childNode.getFirstChild().getNodeValue());
            } else if ("age".equals(childNode.getNodeValue())) {
                boy.setAge(new Integer(childNode.getFirstChild().getNodeValue()));
            }
        }
    }
    list.add(boy);
}
return list;
```

（2）SAX（Simple API for XML）是一种 XML 解析的替代方法。相比于 DOM，SAX 是一种速度更快、更有效的方法。它逐行扫描文档，一边扫描一边解析，而且相比于 DOM，SAX 可以在任意时刻停止解析。但任何事物都有其相反的一面，SAX 的操作比较复杂。

```java
package com.jpy.myapplication;

import java.io.InputStream;
import java.util.ArrayList;
import java.util.List;
import javax.xml.parsers.SAXParser;
import javax.xml.parsers.SAXParserFactory;
import org.xml.sax.Attributes;
import org.xml.sax.SAXException;
```

```java
import org.xml.sax.helpers.DefaultHandler;

public class SAXboyService {
    public List getboys(InputStream inStream) throws Throwable {
        SAXParserFactory factory = SAXParserFactory.newInstance();
        SAXParser parser = factory.newSAXParser();
        boyParse boyParser = new boyParse();
        parser.parse(inStream, boyParser);
        inStream.close();
        return boyParser.getboy();
    }

    private final class boyParse extends DefaultHandler {
        private List list=null;
        Boy boy = null;
        private String tag = null;
        public List getboy() {
            return list;
        }

        @Override
        public void startDocument() throws SAXException {
            list = new ArrayList();
        }

        @Override
        public void startElement(String uri, String localName, String qName, Attributes attributes)
                throws SAXException {
            if ("boy".equals(localName)) {
                boy = new boy();
                boy.setId(new Integer(attributes.getValue(0)));
            }
            tag = localName;//保存元素节点名称
        }

        @Override
        public void endElement(String uri, String localName, String qName) throws SAXException {
            if ("boy".equals(localName)) {
                list.add(boy);
                boy = null;
            }
            tag = null;//结束时,需要清空tag
        }

        @Override
        public void characters(char[] ch, int start, int length) throws SAXException {
```

```
                if (tag != null) {
                    String data = new String(ch, start, length);
                    if ("name".equals(tag)) {
                        boy.setName(data);
                    } else if ("age".equals(tag)) {
                        boy.setAge(new Integer(data));
                    }
                }
            }
        }
    }
}
```

（3）PULL 解析器的运行方式和 SAX 类似，都是基于事件的模式。不同的是，在 PULL 解析过程中，用户需要自己获取产生的事件然后做相应的操作，而不像 SAX 那样由处理器触发一种事件的方法来执行代码。PULL 解析器小巧轻便，解析速度快，简单易用，在 Android 的内核中已经嵌入了 PULL，所以不需要再添加第三方 jar 包来支持 PULL。因此 PULL 非常适合在 Android 移动设备中使用。

```
package com.jpy.myapplication;

import java.io.InputStream;
import java.io.Writer;
import java.util.ArrayList;
import java.util.List;
import org.xmlpull.v1.XmlPullParser;
import org.xmlpull.v1.XmlSerializer;
import android.util.Xml;


public class PullBoyService {
    //保存XML文件
    public static void saveXML(List list, Writer write) throws Throwable {
        XmlSerializer serializer = Xml.newSerializer();
        //序列化
        serializer.setOutput(write);
        //输出流
        serializer.startDocument("UTF-8", true);
        //开始文档
        serializer.startTag(null, "boys");
        //循环去添加
        for (Boy boy : list) {
            serializer.startTag(null, "boy");
            serializer.attribute(null, "id", boy.getId().toString());

            serializer.startTag(null, "name");
            serializer.text(boy.getName());
```

```java
            serializer.endTag(null, "name");
            serializer.startTag(null, "age");
            serializer.text(boy.getAge().toString());

            serializer.endTag(null, "age");
            serializer.endTag(null, "boy");
        }
        serializer.endTag(null, "boys");
        serializer.endDocument();
        write.flush();
        write.close();
    }

    public List getboys(InputStream stream) throws Throwable {
        List list = null;

        Boy boy = null;
        XmlPullParser parser = Xml.newPullParser();
        parser.setInput(stream, "UTF-8");
        int type = parser.getEventType();

        while (type != XmlPullParser.END_DOCUMENT){
            switch (type) {
                case XmlPullParser.START_DOCUMENT:
                    list = new ArrayList();
                    break;
                case XmlPullParser.START_TAG:
                    String name = parser.getName();
                    //获取解析器当前指向的元素名称
                    if ("boy".equals(name)) {
                        boy = new boy();
                        boy.setId(new Integer(parser.getAttributeValue(0)));
                    }
                    if (boy != null) {
                        if ("name".equals(name)) {
                            boy.setName(parser.nextText());

                        }
                        if ("age".equals(name)) {
                            boy.setAge(new Integer(parser.nextText()));
                        }
                    }
                    break;
                case XmlPullParser.END_TAG:
                    if ("boy".equals(parser.getName())) {
                        list.add(boy);
```

```
                    boy = null;
                }
                break;
            }
            type = parser.next();
        }
        return list;
    }
}
```

7.5　问题与讨论

常见的 XML 解析器分别为 SAX 解析器、DOM 解析器和 PULL 解析器，这三者的区别是什么？

项目 8 基于网络通信的天气应用——天气预报（二）

通过前面多个项目的实战，我们已经掌握了 Android 开发技术的核心知识。本项目在项目 2 离线天气预报的基础上实现一个完整的网络天气预报应用。

项目需求描述如下：

（1）在天气预报（一）基础上增加网络通信功能，可实时获取天气信息。
（2）实现根据不同的城市显示相应的天气信息。
（3）根据雨、雪、阴、晴等天气情况分别显示不同的图标。

（1）GridView 的使用方法。
（2）学习使用 HttpClient 工具处理 HTTP 网络通信。
（3）利用 GSON 工具包解析 JSON 数据。
（4）掌握 AsyncTask 的使用方法。
（5）网络图片的处理方法。
（6）较复杂程序的架构设计——网络通信的封装。

8.1 总体设计

8.1.1 总体分析

在项目 2 应用的基础上，增加实时获取在线天气信息数据的功能，数据通过网络从服务器获取并缓存在程序中。

在原有的总体模块、基础架构模块、用户界面模块、数据管理与控制模块的基础上，增加网络通信模块。数据管理与控制模块根据用户界面模块的需求调用网络通信模块，得到城市天气数据并返回到用户界面。

在整个系统中，总体模块控制系统的生命周期；用户界面模块负责显示城市的天气数据、天气状态图标以及各个城市间的显示切换；数据管理与控制模块主要提供数据管理功能，为用户界面模块提供数据，同时可以接收并保存用户界面模块产生的数据；网络通信模块负责发送网络请求并接收网络返回的数据信息。

8.1.2 功能模块框图

根据总体分析结果可以总结出功能模块框图,如图 8-1 所示。

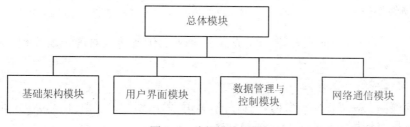

图 8-1 功能模块框图

总体模块的作用主要是生成应用程序的主类,控制应用程序的生命周期;基础架构模块主要提供程序架构、所有 Activity 公用的父类、所有 Activity 公用的方法,包括自定义风格对话框、自定义提示框等功能;数据管理与控制模块主要提供数据获取、数据解析、数据读写、数据组织和数据缓存功能。用户界面模块包括城市天气信息显示、城市管理显示、操作提示等功能;网络通信模块主要负责从服务端获取天气数据信息,数据通过百度提供的天气预报查询接口获取,关于百度数据接口请参考百度网站上的说明文档。

8.1.3 系统流程图

根据总体分析结果及功能模块框图梳理出系统启动的主要流程,如图 8-2 所示。

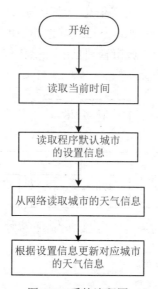

图 8-2 系统流程图

8.1.4 界面设计

同项目 2,具体请参考 2.1.4 节。

8.2 详细设计

8.2.1 模块描述

在系统总体分析及界面布局设计完成后,主要工作就转入对各个功能模块的详细设计阶段。

1. 基础架构模块详细设计

基础架构模块主要提供程序架构、所有 Activity 公用的父类、所有 Activity 公用的方法,包括自定义风格对话框、自定义提示框等功能。

基础架构模块功能如图 8-3 所示。

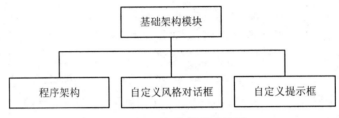

图 8-3 基础架构模块功能图

2. 用户界面模块详细设计

用户界面模块的主要任务是显示天气信息和实现与用户的交互,即当用户点击按键或者屏幕的时候,监听器会去调用相应的处理办法或其他相应的处理模块。

本模块包括城市天气信息显示、城市管理显示、操作提示等功能。

用户界面模块功能如图 8-4 所示。

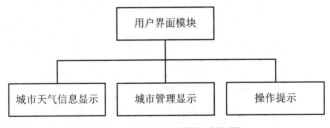

图 8-4 用户界面模块功能图

用户界面模块序列如图 8-5 所示。

3. 数据管理与控制模块详细设计

数据管理与控制模块主要提供数据获取、数据解析、数据读写、数据组织和数据缓存功能。

数据管理与控制模块和用户界面模块可以调用基础架构模块的一些通用方法。数据管理与控制模块为用户界面模块提供数据,同时可以接收并保存用户界面模块产生的数据。

数据管理与控制模块功能图如图 8-6 所示。

项目 8　基于网络通信的天气应用——天气预报（二）

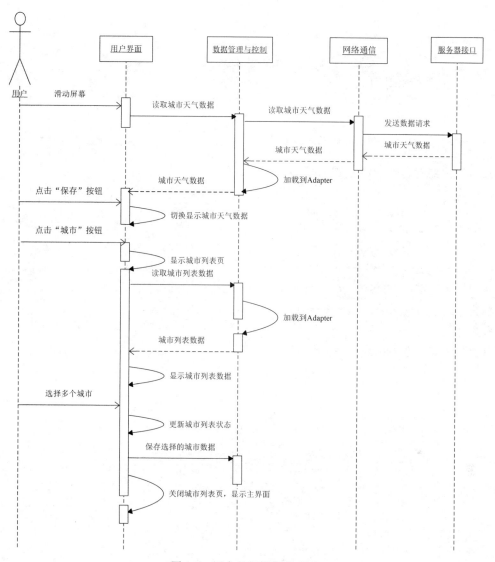

图 8-5　用户界面模块序列图

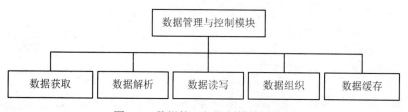

图 8-6　数据管理与控制模块功能图

4. 网络通信模块详细设计

网络通信模块根据用户界面的需求调用访问服务器，接收服务器返回的数据并解析，同时显示到用户界面。

本模块包括发送网络请求、接收网络应答、网络数据解析等功能。

网络通信模块功能图如图 8-7 所示。

图 8-7　网络通信模块功能图

8.2.2　系统包及其资源规划

1. 文件结构

在系统各个模块的实现方式和流程设计完成后，就可以对系统主要的包和资源进行规划，划分的原则主要是保持各个包相互独立，耦合度尽量低。

根据系统功能设计，本系统封装一个基础的 Activity 类，加载各个子页的通用控件，并提供一些基础的实现方法，例如设置进度条、标题等常用的方法，程序中的 Activity 都可继承此基类，继承后就可直接使用基类中封装的基础方法。

系统使用两个 Activity，一个用于显示城市天气信息，一个用于显示城市设置列表。包及其资源结构如图 8-8 所示。

图 8-8　包及其资源结构

2. 命名空间

本示例设置了多个命名空间，分别用来保存用户界面、后台服务的源代码文件，具体说明见表 8-1。

表 8-1　命名空间

命名空间	说明
com.longdong.studio.section9_weather_b	存放与用户界面相关的源代码文件
com.longdong.studio.section9_weather_b.adapter	存放与页面适配器相关的源代码文件
com.longdong.studio.section9_weather_b.model	存放与数据结构基础类相关的源代码文件
com.longdong.studio.section9_weather_b.network	存放与网络通信相关的源代码文件

3. 源代码文件

源代码文件及说明见表 8-2。

表 8-2　源代码文件

包名称	文件名	说明
com.longdong.studio.section9_weather_b	BaseActivity.java	页面的基类
	CitySelectionActivity.java	选择城市页的 Activity
	MainActivity.java	主界面页的 Activity
com.longdong.studio.section9_weather_b.adapter	CityAdapter.java	城市列表的 Adapter
	MyPagerAdapter.java	主界面切换城市用 ViewPager 的 Adapter
	WeatherInfoAdapter.java	主界面显示未来几天天气情况的 ListView 的 Adapter
com.longdong.studio.section9_weather_b.model	CityInfo.java	城市信息数据操作基础类
	ResultModel.java	结果状态数据类
	WeatherInfo.java	保存天气信息数据
	WeatherInfoWrapper.java	天气信息网络数据转换
com.longdong.studio.section9_weather_b.network	Common.java	通用变量和方法定义，属于基础模块
	HttpTask.java	网络通信任务封装
	HttpUtils.java	网络请求应答方法封装

4. 资源文件

Android 的资源文件保存在 /res 的子目录中。

- /res/drawable 目录：保存的是图像文件。
- /res/layout 目录：保存的是布局文件。
- /res/values 目录：保存的是用来定义字符串和颜色的文件。

资源文件及说明见表 8-3。

表 8-3 资源文件

资源目录	文件	说明
drawable	baoyu.png	暴雨的图标
	dayu.png	大雨的图标
	duoyun.png	多云的图标
	ic_launcher.png	程序图标文件
	menuback.jpg	主界面背景图
	qing.png	晴的图标
	xiaoyu.png	小雨的图标
	yin.png	阴的图标
	zhenyu.png	阵雨的图标
	zhongyu.png	中雨的图标
layout	activity_base.xml	基础页的布局
	activity_main.xml	主界面的布局
	city_item.xml	城市列表的布局
	weather_info_item.xml	列表行的布局
values	dimens.xml	保存尺寸的 XML 文件
	strings.xml	保存字符串的 XML 文件
	styles.xml	保存样式的 XML 文件

8.2.3 主要方法流程设计

查看天气流程图如图 8-9 所示。

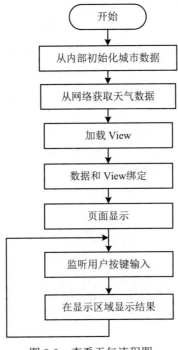

图 8-9 查看天气流程图

8.3 代码实现

8.3.1 显示界面布局

同项目 2，具体请参考 2.3.1 节。

8.3.2 控件设计实现

同项目 2，具体请参考 2.3.2 节。

8.3.3 天气预报接口方法实现

Android 4.0 以上版本强制要求不能在主线程中执行耗时的网络操作，网络操作需要使用 Thread+Handler 或 AsyncTask，这里将介绍 AsyncTask 的使用方法。

1. 添加 HttpTask.java

```java
/**
 * Params：task执行excute方法时传入的可变参数，同时被传入doInBackground
 * Progress：执行进度相关的可变长参数，由publishProgress（在doInBackground中）方法传入，
 *           在onProgressUpdate方法中获取
 * Result：doInBackground返回的结果，同时被传入
 */
public class HttpTask extends AsyncTask<RequestData,Integer, ResposneBundle> {
    private BaseActivity mContext;
    public interface TaskResultListener {
        void result(ResposneBundle b);
    }
    TaskResultListener taskResult;
    public HttpTask(BaseActivity context,TaskResultListener taskResult){
        mContext = context;
        this.taskResult = taskResult;
    }
    protected void onPreExecute() {
        mContext.showProgress("加载中");
    }
    protected ResposneBundle doInBackground(RequestData... params) {
        publishProgress(1);
        return HttpUtils.excuteRequest(params[0]);
    }
    /**
     * 后台线程的任务执行完毕，Result即为doInBackground的执行结果
     */
    protected void onPostExecute(ResposneBundle result) {
        mContext.dismissProgress();
        if(result != null){
            taskResult.result(result);
```

```java
                }else{
                }
            }
            protected void onProgressUpdate(Integer... values) {
                //textView.setText("图片尺寸："+values[2]+" \r\n已下载大小："+values[1]+" \r\n当前下载
                进度："+values[0]+"%");
            }
        }
```

2. 调用使用

```java
//下载天气图标，使用文件缓存方式。如果已缓存在本地，则不再进行网络请求，否则直接从网络获取图片
private void loadIcon(final ImageView iconImg,final WeatherInfo info){
            //获取BaseActivity
            BaseActivity act = (BaseActivity) iconImg.getContext();
            RequestData data = new RequestData();
            int hour = Calendar.getInstance().get(Calendar.HOUR_OF_DAY);
            Log.e("hour",hour+"");
            if(hour>6&&hour<=18){
                        data.uri   = info.getDayPictureUrl();
                        info.filePath = act.getDir("icon\\day", Context.MODE_PRIVATE);
            }else{
                        data.uri   = info.getNightPictureUrl();
                        info.filePath = act.getDir("icon\\night", Context.MODE_PRIVATE);
            }

            String fileName = data.uri.substring(data.uri.lastIndexOf("/"),data.uri.length());
            Log.e("fileName",fileName);
            info.filePath = new File(info.filePath,fileName);
            Log.e("Full path",info.filePath.getAbsolutePath());
            //如果本地缓存存在，则直接显示本地缓存中的图片，否则请求网络中的图片
            if(info.filePath.exists()){
                        Log.e("提示","文件存在，从本地获取");
                        info.bmp = BitmapFactory.decodeFile(info.filePath.getAbsolutePath());
                        iconImg.setImageBitmap(info.bmp);
                //去除Bitmap的强引用
                info.bmp = null;
            }else{
                Log.e("提示","文件不存在，从网络请求");
                act.startHttpTask(new TaskResultListener() {
                    @Override
                    public void result(ResposneBundle b) {
                        Log.e("RRR",new String(b.result));
                        info.bmp =   BitmapFactory.decodeByteArray(b.result, 0, b.result.length);
                        if(info.bmp==null){
                            Log.e("出错了","天气图标下载失败！");
                            return;}
                        try {
```

```
                    FileOutputStream out = new FileOutputStream(info.filePath);
                    info.bmp.compress(Bitmap.CompressFormat.PNG, 100, out);
                } catch (FileNotFoundException e) {
                    e.printStackTrace();
                }

                iconImg.setImageBitmap(info.bmp);
                //去除Bitmap的强引用
                info.bmp = null;

            }
        }, data);
    }
}
//调用函数中用到了BaseActivity中的startHttpTask进行网络请求
public void startHttpTask(TaskResultListener l, RequestData data) {
    task = new HttpTask(this, l);
    task.execute(data);
}
```

8.4 关键知识点解析

8.4.1. 在程序中使用天气预报接口

国家气象局提供了三种数据形式的天气预报接口，我们可以直接使用，具体如下：

- http://www.weather.com.cn/data/sk/城市ID.html
- http://www.weather.com.cn/data/cityinfo/城市ID.html
- http://m.weather.com.cn/data/城市ID.html

北京的ID为101010100，以北京为例的JSON格式如下：

第一个网址提供的JSON数据格式：

```
{
    "weatherinfo": {
        "city": "北京",
        "cityid": "101010100",
        "temp": "27.9",
        "WD": "尪",
        "WS": " ",
        "SD": "28%",
        "AP": "1002hPa",
        "njd": "",
        "WSE": "<3",
        "time": "17:55",
        "sm": "2.1",
        "isRadar": "1",
```

```
            "Radar": "JC_RADAR_AZ9010_JB"
        }
}
```
第二个网址提供的 JSON 数据格式：

{"weatherinfo":{"city":"北京","cityid":"101010100","temp1":"18 ","temp2":"31 ","weather":"","img1":"n1.gif","img2":"d2.gif","ptime":"18:00"}}

第三个网址提供的 JSON 数据格式较为全面，具体如下：

```
{
    "weatherinfo": {
        "city": "北京",
        "city_en": "beijing",
        "date_y": "2014年3月4日",
        "date": "",
        "week": "星期二",
        "fchh": "11",
        "cityid": "101010100",
        "temp1": "8℃~-3℃",
        "temp2": "8℃~-3℃",
        "temp3": "7℃~-3℃",
        "temp4": "8℃~-1℃",
        "temp5": "10℃~1℃",
        "temp6": "10℃~2℃",
        "tempF1": "46.4℉~26.6℉",
        "tempF2": "46.4℉~26.6℉",
        "tempF3": "44.6℉~26.6℉",
        "tempF4": "46.4℉~30.2℉",
        "tempF5": "50℉~33.8℉",
        "tempF6": "50℉~35.6℉",
        "weather1": "晴",
        "weather2": "晴",
        "weather3": "晴",
        "weather4": "晴转多云",
        "weather5": "多云",
        "weather6": "多云",
        "img1": "0",
        "img2": "99",
        "img3": "0",
        "img4": "99",
        "img5": "0",
        "img6": "99",
        "img7": "0",
        "img8": "1",
        "img9": "1",
        "img10": "99",
        "img11": "1",
        "img12": "99",
```

```
"img_single": "0",
"img_title1": "晴",
"img_title2": "晴",
"img_title3": "晴",
"img_title4": "晴",
"img_title5": "晴",
"img_title6": "晴",
"img_title7": "晴",
"img_title8": "多云",
"img_title9": "多云",
"img_title10": "多云",
"img_title11": "多云",
"img_title12": "多云",
"img_title_single": "晴",
"wind1": "北风4~5级转微风",
"wind2": "微风",
"wind3": "微风",
"wind4": "微风",
"wind5": "微风",
"wind6": "微风",
"fx1": "北风",
"fx2": "微风",
"fl1": "4~5级转小于3级",
"fl2": "小于3级",
"fl3": "小于3级",
"fl4": "小于3级",
"fl5": "小于3级",
"fl6": "小于3级",
"index": "寒冷",
"index_d": "天气寒冷，建议着厚羽绒服、毛皮大衣加厚毛衣等隆冬服装。年老体弱者尤其要注意保暖防冻。",
"index48": "冷",
"index48_d": "天气冷，建议着棉服、羽绒服、皮夹克加羊毛衫等冬季服装。年老体弱者宜着厚棉衣、冬大衣或厚羽绒服。",
"index_uv": "中等",
"index48_uv": "中等",
"index_xc": "较适宜",
"index_tr": "一般",
"index_co": "较舒适",
"st1": "7",
"st2": "-3",
"st3": "8",
"st4": "0",
"st5": "7",
"st6": "-1",
"index_cl": "较不宜",
```

```
            "index_ls": "基本适宜",
            "index_ag": "易发"
    }
}
```

8.4.2 采用 MQTT 协议实现 Android 推送

1. 客户端

客户端具体代码如下：

```java
import android.content.Context;
import android.content.Intent;
import android.content.SharedPreferences;
import android.media.MediaPlayer;
import android.media.Ringtone;
import android.media.RingtoneManager;
import android.net.Uri;
import android.util.Log;

import com.alibaba.fastjson.JSONObject;
import com.google.protobuf.nano.CodedOutputByteBufferNano;

import cn.tencent.mars.sample.proto.Message;

import org.eclipse.paho.android.service.MqttAndroidClient;
import org.eclipse.paho.android.service.MqttInit;
import org.eclipse.paho.client.mqttv3.IMqttActionListener;
import org.eclipse.paho.client.mqttv3.IMqttDeliveryToken;
import org.eclipse.paho.client.mqttv3.IMqttMessageListener;
import org.eclipse.paho.client.mqttv3.IMqttToken;
import org.eclipse.paho.client.mqttv3.MqttCallbackExtended;
import org.eclipse.paho.client.mqttv3.MqttConnectOptions;
import org.eclipse.paho.client.mqttv3.MqttException;
import org.eclipse.paho.client.mqttv3.MqttMessage;

import java.io.IOException;
import java.security.KeyManagementException;
import java.security.KeyStore;
import java.security.KeyStoreException;
import java.security.NoSuchAlgorithmException;
import java.security.SecureRandom;
import java.security.cert.CertificateException;
import java.util.HashMap;
import java.util.List;
import java.util.Map;

import javax.net.ssl.SSLContext;
import javax.net.ssl.SSLSocketFactory;
```

```java
import javax.net.ssl.TrustManager;
import javax.net.ssl.TrustManagerFactory;

public class MqttManager {

    MqttAndroidClient mqttAndroidClient = null;

    String clientId = "";
    String userId = "";
    Context mContext;
    MediaPlayer mMediaPlayer = new MediaPlayer();
    static final int HAND_GET_USER_INFO = 10001;

    /**
     * MQTT管理构造方法
     *
     * @param context
     * @param cId          登录账号的clientID
     * @param userId       登录账号的会话ID
     */
    public MqttManager(Context context, String cId, String userId) {
        mContext = context;
        clientId = cId;
        this.userId = userId;

        mqttAndroidClient = new MqttAndroidClient();
        mqttAndroidClient.setCallback(new MqttCallbackExtended() {
            @Override
            public void connectComplete(boolean reconnect, String serverURI) {

                if (reconnect) {              //重新连接
                    subscribeToTopic();       //订阅自己
                } else {
                    addToLog("Connected to: " + serverURI);
                }
            }

            @Override
            public void connectionLost(Throwable cause) {
                addToLog("The Connection was lost.");
            }

            @Override
            public void messageArrived(String topic, MqttMessage message) throws Exception {
                addToLog("messageArrived message: " + new String(message.getPayload()));
                opMessageArrived(topic, message);
```

```java
            }

            @Override
            public void deliveryComplete(IMqttDeliveryToken token) {

            }
        });
    }

    //是否被移除释放了
    public boolean isRelease()
    {
        if(mqttAndroidClient ==null ||mqttAndroidClient==null)
            return true;
        return false;
    }
    /**
     * 连接MQTT
     */
    public void connect() {
        MqttConnectOptions mqttConnectOptions = new MqttConnectOptions();
        mqttConnectOptions.setAutomaticReconnect(true);
        mqttConnectOptions.setCleanSession(false);

        mqttConnectOptions.setConnectionTimeout(5);
        mqttConnectOptions.setKeepAliveInterval(5);

        try {

            //连接MQTT
            mqttAndroidClient.connect(mqttConnectOptions, null, new IMqttActionListener() {
                @Override
                public void onSuccess(IMqttToken asyncActionToken) {
                    subscribeToTopic();
                }

                @Override
                public void onFailure(IMqttToken asyncActionToken, Throwable exception) {
                    addToLog("Failed to connect to: " + Global.mqttserverUri);
                }
            });

        } catch (MqttException ex) {
            ex.printStackTrace();
        } catch (IOException e) {
            e.printStackTrace();
```

```java
        }
    }

/**
 * 订阅自己接收消息
 */
public void subscribeToTopic() {

    try {

        mqttAndroidClient.subscribe(userId, 2, null, new IMqttActionListener() {
            @Override
            public void onSuccess(IMqttToken asyncActionToken) {
                addToLog("Subscribed!" + userId);
                online();
            }

            @Override
            public void onFailure(IMqttToken asyncActionToken, Throwable exception) {
                addToLog("Failed to subscribe");
            }
        });

        // THIS DOES NOT WORK!
        mqttAndroidClient.subscribe(userId, 2, new IMqttMessageListener() {
            @Override
            public void messageArrived(String topic, MqttMessage message) throws Exception {
                // message Arrived!
                opMessageArrived(topic, message);
            }
        });

    } catch (MqttException ex) {
        addToLog("Exception whilst subscribing");
        ex.printStackTrace();
    }
}


/**
 * 订阅取消自己接收消息
 */
public void unSubscribe(String userId) {
    try {
        offline();
        mqttAndroidClient.unsubscribe(userId);
```

```java
            } catch (MqttException e) {
                e.printStackTrace();
            }
        }

        /**
         * 处理接收到的消息
         *
         * @param topic
         * @param message
         */
        void opMessageArrived(String topic, MqttMessage message) {
            try {
                Message.ExchangeMessage msg = Message.ExchangeMessage.parseFrom(message.getPayload());

            } catch (Exception ex) {
                ex.printStackTrace();
            }
        }

        //通知前台刷新
        public void saveMsg(String msgid, String topic, String status) {
            //更新消息状态
            MsgRecordDao msgRecordDao = new MsgRecordDao(mContext);
            msgRecordDao.setMsgStatus(msgid, status);

            Intent intent = new Intent();
            intent.setAction(Global.ACTION_RECV_MSG);
            mContext.sendBroadcast(intent);
            ChatDataCore.getInstance().setData(topic,msgid,status);
        }

        public void refresh(String msgid, String topic, String status) {
            //更新消息状态
            MsgRecordDao msgRecordDao = new MsgRecordDao(mContext);
            msgRecordDao.setMsgStatus(msgid, status);

            Intent intent = new Intent();
            intent.setAction(Global.ACTION_RECV_MSG);
            mContext.sendBroadcast(intent);
            ChatDataCore.getInstance().setData(topic,msgid,status);
        }

        //通知刷新
        public void refreshStatus(String msgid, String topic, String status) {
            //更新消息状态
```

```java
        MsgRecordDao msgRecordDao = new MsgRecordDao(mContext);
        msgRecordDao.setMsgStatus(msgid, status);
        ChatDataCore.getInstance().setData(topic,msgid,status);
    }

    //通知刷新
    public void refreshGroupStatus(String msgid, String topic, String status, String sendid) {
        //更新消息状态
        MsgRecordDao msgRecordDao = new MsgRecordDao(mContext);
        msgRecordDao.AddUsers(msgid, sendid);
        ChatDataCore.getInstance().setData(topic,msgid,status);
    }

    /**
     * 发送消息
     *
     * @param msg  消息内容
     */
    public void publishMessage(final Message.ExchangeMessage msg) {
//DebugLog.LogE(msg.toString());
        if(msg.wordPayload==null)
        {
            Message.WordPayload txt2 = new Message.WordPayload();
            txt2.word = "";
            msg.wordPayload = txt2;
        }
        publishMessage(msg, new IMqttActionListener() {
            @Override
            public void onSuccess(IMqttToken asyncActionToken) {
                //更新成功记录
                //if(msg.payloadType!=2)
                //图片单独提示成功
                //通知前台刷新

                saveMsg(msg.id, msg.receiverId, Global.CHAT_STATUS_UNREAD);
                addToLog("Message Published");
            }

            @Override
            public void onFailure(IMqttToken asyncActionToken, Throwable exception) {
                //更新失败记录
                //通知前台刷新
                saveMsg(msg.id, msg.receiverId, Global.CHAT_STATUS_FAIL);
                addToLog("Message onFailure");
            }
        });
```

```java
    }
    public void publishMessage(final Message.ExchangeMessage msg, IMqttActionListener iMqttActionListener) {

        try {
            if (mqttAndroidClient == null) {
                return;
            }
            if (msg == null) {
                return;
            }
            int length = msg.getSerializedSize();
            byte[] flatArray = new byte[length];
            final CodedOutputByteBufferNano output = CodedOutputByteBufferNano.newInstance(flatArray);
            msg.writeTo(output);

            MqttMessage message = new MqttMessage();
            message.setPayload(flatArray);
            message.setQos(2);
            message.setRetained(false);
            mqttAndroidClient.publish(Global.EXCHANG_EMESSAGE_TOPIC, message, null,
                    iMqttActionListener);
            if (!mqttAndroidClient.isConnected()) {
                addToLog(" messages in buffer.");
            }

        } catch (MqttException e) {
            addToLog("Error Publishing: " + e.getMessage());
            e.printStackTrace();
        } catch (IOException e) {
            e.printStackTrace();
        }
    }

    public void online() {
        try {
            Message.FunctionMessage fmsg = new Message.FunctionMessage();
            fmsg.id = Device.getUUID();
            fmsg.clientId = clientId;
            fmsg.messageType = 0;
            fmsg.userId = userId;
            byte[] flatArray = new byte[fmsg.getSerializedSize()];
            final CodedOutputByteBufferNano output = CodedOutputByteBufferNano.newInstance(flatArray);
            fmsg.writeTo(output);

            MqttMessage message = new MqttMessage();
            message.setPayload(flatArray);
```

```java
            mqttAndroidClient.publish(Global.FUNCTIONMESAGETOPIC, message);
        } catch (MqttException e) {
            System.err.println("Error online: " + e.getMessage());
            e.printStackTrace();
        } catch (IOException e) {
            e.printStackTrace();
        }
    }

    public void DisConnect() {
        if (mqttAndroidClient != null)
            addToLog("close connect: " + userId);
        try {
            mqttAndroidClient.disconnect();
        } catch (MqttException e) {
            e.printStackTrace();
        }
    }
}
```

2．服务端

服务端具体代码如下：

```java
package com.jpy.myapplication;

import android.app.Service;

public class Server extends Service {
    JButton button;
    JPanel panel;
    private MqttClient client;
    private String host = "tcp://127.0.0.1:1883";

    private String userName = "test";
    private String passWord = "test";
    private MqttTopic topic;
    private MqttMessage message;

    private String myTopic = "android/topic";

    public Server() {

        try {
            client = new MqttClient(host, "Server",
                    new MemoryPersistence());
            connect();
        } catch (Exception e) {
            e.printStackTrace();
```

```java
        }

        Container container = this.getContentPane();
        panel = new JPanel();
        button = new JButton("发布话题");
        button.addActionListener(new WifiP2pManager.ActionListener() {

            @Override
            public void actionPerformed(ActionEvent ae) {
                try {
                    MqttDeliveryToken token = topic.publish(message);
                    token.waitForCompletion();

                } catch (Exception e) {
                    e.printStackTrace();
                }
            }
        });
        panel.add(button);
        container.add(panel, " ");

    }

    private void connect() {

        MqttConnectOptions options = new MqttConnectOptions();
        options.setCleanSession(false);
        options.setUserName(userName);
        options.setPassword(passWord.toCharArray());
        // 设置超时时间
        options.setConnectionTimeout(10);
        // 设置会话心跳时间
        options.setKeepAliveInterval(20);
        try {
            client.setCallback(new MqttCallback() {

                @Override
                public void connectionLost(Throwable cause) {

                }

                @Override
                public void deliveryComplete(IMqttDeliveryToken token) {

                }
```

```java
            @Override
            public void messageArrived(String topic, MqttMessage arg1)
                    throws Exception {

            }
        });

        topic = client.getTopic(myTopic);

        message = new MqttMessage();
        message.setQos(2);
        message.setRetained(true);

        message.setPayload("消息".getBytes());

        client.connect(options);
    } catch (Exception e) {
        e.printStackTrace();
    }

}

public static void main(String[] args) {
    Server s = new Server();
    s.setDefaultCloseOperation(JFrame.EXIT_ON_CLOSE);
    s.setSize(300, 300);
    s.setLocationRelativeTo(null);
    s.setVisible(true);
}
}
```

8.5 问题与讨论

在 Android 中，常见的消息推送方式有轮询（Pull）、SMS（Push）和持久连接（Push），简述这三者间的区别。

项目 9 商务通讯录

本项目为商务人士开发一款基于智能手机通讯录增强功能的应用,该应用具有通讯录备份和同步功能、通话历史记录统计功能、附加商务名片编辑功能、来电时商务名片及相关信息即时显示功能。

项目需求描述如下:
(1)用户注册和登录模块。
(2)手机联系人模块,功能包括展示联系人列表、详细信息、拨打电话等。
(3)设计合理的算法,尽可能减少数据传输量,同步服务器与本地通讯录,实现云端记录覆盖手机端记录和手机端记录覆盖云端记录的功能。
(4)可以在手机端为联系人附加商务名片等额外信息。
(5)当具有额外信息的联系人来电时,屏幕显示该联系人的商务名片。
(6)支持输入电话号码或中文名字查找联系人,在输入过程中,匹配结果应实时进行过滤。
(7)可以查看手机中所有联系人通话的本地历史记录。
(8)联系人查找功能支持拼音首字母或拼音模糊查询。
(9)使用本地数据库判断来电是否为黑名单号码,并给用户提醒。

(1)学习访问手机通讯录。
(2)学习拦截来电。

9.1 总体设计

9.1.1 总体分析

该应用通过备份手机通讯录和自动与服务器同步的机制保证用户通讯录的安全与完整,当用户使用同一账号在不同手机登录应用程序时,通过对比计算确保不同手机上的通讯录信息一致。

9.1.2 功能模块框图

根据总体分析结果可以总结出功能模块框图,如图9-1所示。

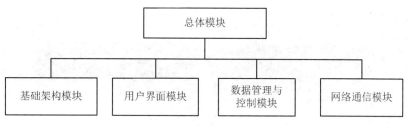

图 9-1　功能模块框图

总体模块的作用主要是生成应用程序的主类，控制应用程序的生命周期；基础架构模块主要提供程序架构、所有 Activity 公用的父类、所有 Activity 公用的方法，包括自定义风格对话框、自定义提示框等功能；数据管理与控制模块主要提供数据获取、数据解析、数据读写、数据组织和数据缓存功能；用户界面模块包括通讯录列表显示、备份还原通讯录、来电黑名单设置、操作提示等功能；网络通信模块主要负责从服务端获取联系人列表和通讯录备份。

9.1.3　系统流程图

根据总体分析结果及功能模块框图梳理出系统启动的主要流程，如图 9-2 所示。

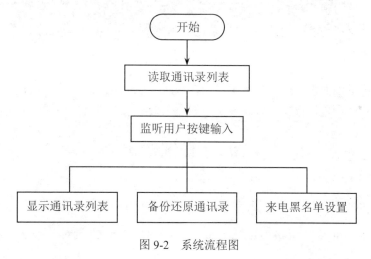

图 9-2　系统流程图

9.1.4　界面设计

本应用是显示通讯录的应用，主要实现通讯录列表显示、备份还原通讯录和来电黑名单设置。

根据程序功能需求可以规划出软件的主要界面，如下：

（1）通讯录列表：启动应用程序后显示通讯录列表。

（2）备份还原通讯录：点击底部的"设置"按钮切换页面后，点击"备份还原通讯录"进入。

（3）来电黑名单设置：点击底部的"设置"按钮切换页面后，点击"来电黑名单设置"进入。

系统主界面如图 9-3 所示。

图 9-3　系统主界面

从图 9-3 中可以很直观地看到，主界面包含两个选项卡：通讯录和设置，"通讯录"选项卡包含姓名和电话，"设置"选项卡包含备份还原通讯录和来电黑名单设置两个功能。

9.2　详细设计

9.2.1　模块描述

在系统总体分析及界面布局设计完成后，主要工作就转入对各个功能模块的详细设计阶段。

1. 基础架构模块详细设计

基础架构模块主要提供程序架构、所有 Activity 公用的父类、所有 Activity 公用的方法，包括自定义风格对话框、自定义提示框等功能。

基础架构模块功能如图 9-4 所示。

图 9-4　基础架构模块功能图

2. 用户界面模块详细设计

用户界面模块的主要任务是显示通讯录信息和实现与用户的交互，即当用户点击按键或者屏幕的时候，监听器会去调用相应的处理办法或其他相应的处理模块。

本模块包括通讯录显示、通讯录还原备份、来电黑名单设置等功能。

用户界面模块功能如图 9-5 所示。

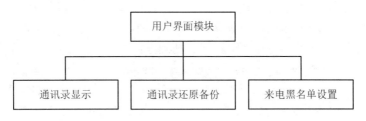

图 9-5　用户界面模块功能图

3. 数据管理与控制模块详细设计

数据管理与控制模块主要提供数据获取、数据解析、数据读写、数据组织和数据缓存功能。

数据管理与控制模块和用户界面模块可以调用基础架构模块的一些通用方法。数据管理与控制模块为用户界面模块提供数据，同时可以接收并保存用户界面模块产生的数据。

数据管理与控制模块功能如图 9-6 所示。

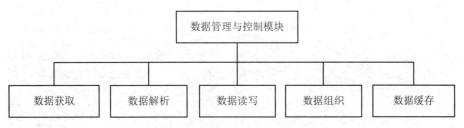

图 9-6　数据管理与控制模块功能图

4. 网络通信模块详细设计

网络通信模块根据用户界面的需求调用访问服务器，接收服务器返回的数据并解析，同时显示到用户界面。

本模块包括发送网络请求、接收网络应答、网络数据解析等功能。

网络通信模块功能如图 9-7 所示。

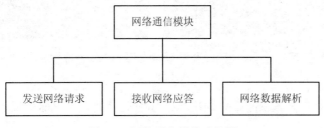

图 9-7　网络通信模块功能图

9.2.2 系统包及其资源规划

1. 文件结构

在系统各个模块的实现方式和流程设计完成后，就可以对系统主要的包和资源进行规划，划分的原则主要是保持各个包相互独立，耦合度尽量低。

根据系统功能设计，本系统封装一个基础的 Activity 类，加载各个子页的通用控件，并提供一些基础的实现方法，例如设置进度条、标题等常用的方法，程序中的 Activity 都可继承此基类，继承后就可直接使用基类中封装的基础方法。

系统使用两个 Activity 和两个 Fragment，一个 Fragment 用于通讯录信息，一个 Fragment 用于通讯录设置，一个 Activity 显示通讯录还原备份信息，一个 Activity 显示来电黑名单。包及其资源结构如图 9-8 所示。

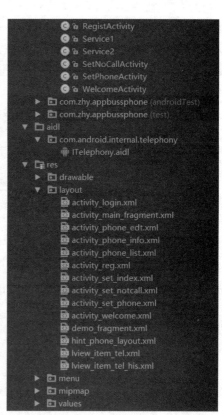

图 9-8 包及其资源结构

2. 命名空间

本示例设置了多个命名空间，分别用来保存用户界面、后台服务的源代码文件，具体说明见表 9-1。

表 9-1 命名空间

命名空间	说明
com.zhy.appbussphone.bean	存放与通讯录用户信息相关的源代码文件
com.zhy.appbussphone.ormlite	存放与数据库相关的源代码文件
com.zhy.appbussphone.rewirte_view	存放与自定义控件相关的源代码文件
com.zhy.appbussphone.tools	存放与工具类相关的源代码文件
com.zhy.appbussphone	存放与页面视图相关的源代码文件

3. 源代码文件

源代码文件及说明见表 9-2。

表 9-2 源代码文件

包名称	文件名	说明
com.zhy.appbussphone.bean	CallInfo.java	通讯录用户信息类
com.zhy.appbussphone.ormlite	TelBook.java	电话信息类
	DaoTelBook.java	电话信息数据库操作类
	DatabaseHelper.java	数据库类
com.zhy.appbussphone.rewirte_view	ClearEditText.java	自定义编辑框类
	CountDownButtonHelper.java	自定义倒计时按钮类
	EditTextField.java	自定义带有删除按钮编辑框类
	ListViewForScrollView.java	自定义 ListView 视图类
com.zhy.appbussphone.tools	StringUtilsZhy.java	字符串工具类
	TelUtil.java	电话工具类
	Tools.java	自定义工具栏高度类
com.zhy.appbussphone	MainFragmentActivity.java	主页 FragmentActivity
	WelcomeActivity.java	欢迎页 Activity
	SetPhoneActivity.java	设置通讯录备份还原 Activity
	SetNoCallActivity.java	来电黑名单 Activity

4. 资源文件

Android 的资源文件保存在/res 的子目录中。

- /res/drawable 目录：保存的是图像文件。
- /res/layout 目录：保存的是布局文件。
- /res/values 目录：保存的是用来定义字符串和颜色的文件。

资源文件及说明见表 9-3。

9.2.3 主要方法流程设计

更新通讯录流程图如图 9-9 所示。

表 9-3 资源文件

资源目录	文件	说明
drawable	splash.png	欢迎页背景
	back_ltr..png	左箭头图标
	back_rtr.png	右箭头图标
	ic_launcher.png	程序图标文件
	bg.jpg	主界面背景图
	chats_green.png	通讯录 tab 的图标
	contacts.png	设置的图标
	set_index_syc.png	备份还原的图标
	set_index_nocall.png	来电黑名单的图标
	tel_find.png	搜索图标
layout	activity_main_fragment.xml	主界面的布局
	activity_set_phone.xml	设置通讯录备份还原页面的布局
	activity_set_nocall.xml	设置来电黑名单页面的布局
	activity_welcome.xml	欢迎页的布局
values	dimens.xml	保存尺寸的 XML 文件
	attrs.xml	保存样式的 XML 文件
	colors.xml	保存颜色的 XML 文件
	strings.xml	保存字符串的 XML 文件
	styles.xml	保存样式的 XML 文件

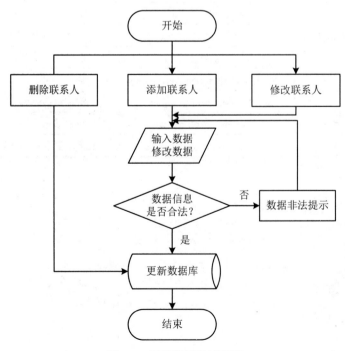

图 9-9 更新通讯录流程图

9.3 代码实现

9.3.1 显示界面布局

1. 系统主界面

系统主界面是进入系统后显示的联系人列表和设置界面，该界面包括一个 ViewPager 控件和两个 Fragment。其中一个 Fragment 中嵌入了一个 ListView、两个 EditText、一个 ImageView，每个 ListView 条目中包含一个 ImageView 和两个 TextView；另一个 Fragment 中嵌入了 4 个 ImageView 和两个 TextView。系统主界面如图 9-10 所示。

图 9-10　系统主界面

2. 联系人详情界面

联系人详情界面用于显示联系人详细信息，可以进行联系人详情编辑和查看。该界面包括十个 TextView 和一个 ImageView，如图 9-11 所示。

3. 还原备份通讯录界面

还原备份通讯录界面可以进行通讯录的备份、还原和同步，该界面包括三个 Button，如图 9-12 所示。

4. 来电黑名单设置界面

来电黑名单设置界面用于设置来电黑名单，该界面包括一个 Button 和一个 EditText，如图 9-13 所示。

图 9-11　联系人详情界面

图 9-12　还原备份通讯录界面

图 9-13　来电黑名单设置界面

9.3.2　控件设计实现

EditText 带有删除按钮，代码如下：

```
public class ClearEditText extends AppCompatEditText implements OnFocusChangeListener, TextWatcher {

    /**
     * 删除按钮的引用
     */
    private Drawable mClearDrawable;
    /**
```

     * 控件是否有焦点
     */
    private boolean hasFoucs;

    public ClearEditText(Context context) {
        this(context, null);
    }

    public ClearEditText(Context context, AttributeSet attrs) {
        //这里构造方法也很重要，不加这个很多属性不能在XML里面定义
        this(context, attrs, android.R.attr.editTextStyle);
    }

    public ClearEditText(Context context, AttributeSet attrs, int defStyle) {
        super(context, attrs, defStyle);
        init();
    }

    private void init() {
        //获取EditText的DrawableRight，假如没有设置我们就使用默认的图片
        mClearDrawable = getCompoundDrawables()[2];
        if (mClearDrawable == null) {
            mClearDrawable = getResources().getDrawable(R.drawable.edittext_delete_selector);
        }

        mClearDrawable.setBounds(0, 0, mClearDrawable.getIntrinsicWidth(),
            mClearDrawable.getIntrinsicHeight());
        //默认设置隐藏图标
        setClearIconVisible(false);
        //设置焦点改变的监听
        setOnFocusChangeListener(this);
        //设置输入框里内容发生改变的监听
        addTextChangedListener(this);
    }

    /**
     *因为不能直接给EditText设置点击事件，所以用记住按下的位置来模拟点击事件,
     *用输入框的onTouchEvent()方法来模拟。
     *当按下的位置在图标的左边线（EditText的宽度－图标到控件右边的间距－图标的宽度）和图标的
     *右边线（EditText的宽度－图标到控件右边的间距）之间时，就算点击了图标，竖直方向没有考虑。
     */
    @Override
    public boolean onTouchEvent(MotionEvent event) {
        if (event.getAction() == MotionEvent.ACTION_UP) {
            if (getCompoundDrawables()[2] != null) {
                boolean touchable = event.getX() > (getWidth() - getTotalPaddingRight()) && (event.getX()
                    < ((getWidth() - getPaddingRight()))));

```java
                if (touchable) {
                    this.setText("");
                }
            }
        }
        return super.onTouchEvent(event);
    }

    /**
     * 当ClearEditText焦点发生变化的时候，判断里面字符串长度，设置清除图标的显示与隐藏
     */
    @Override
    public void onFocusChange(View v, boolean hasFocus) {
        this.hasFoucs = hasFocus;
        if (hasFocus) {
            setClearIconVisible(getText().length() > 0);
        } else {
            setClearIconVisible(false);
        }
    }




    /**
     * 设置清除图标的显示与隐藏，调用setCompoundDrawables为EditText绘制上去
     * @param visible
     */
    protected void setClearIconVisible(boolean visible) {
        Drawable right = visible ? mClearDrawable : null;
        setCompoundDrawables(getCompoundDrawables()[0],getCompoundDrawables()[1], right,
            getCompoundDrawables()[3]);
    }

    /**
     * 当输入框里内容发生变化的时候回调的方法
     */
    @Override
    public void onTextChanged(CharSequence s, int start, int count,int after) {
        if(hasFoucs){
            setClearIconVisible(s.length() > 0);
        }
    }

    @Override
    public void beforeTextChanged(CharSequence s, int start, int count,int after) {

    }
```

```java
    @Override
    public void afterTextChanged(Editable s) {

    }
}
```

9.3.3 监听手机来电服务

来电监听服务使用的是 PhoneStateListener 类,使用方式:将 PhoneStateListener 对象注册到系统电话管理服务(TelephonyManager)中去,然后通过 PhoneStateListener 的回调方法 onCallStateChanged()实现来电的监听。

注册监听代码如下:

```java
private void registerPhoneStateListener() {
    MyPhoneStateListener psListener = new MyPhoneStateListener (this);
    TelephonyManager tManager = (TelephonyManager) getSystemService(Context.TELEPHONY_SERVICE);
    if (tManager != null) {
        tManager .listen(psListener , PhoneStateListener.LISTEN_CALL_STATE);
    }
}
```

通过 PhoneStateListener 的 onCallStateChanged()方法监听来电状态,代码如下:

```java
import android.content.Context;
import android.telephony.PhoneStateListener;
import android.telephony.ServiceState;
import android.telephony.TelephonyManager;

public class MyPhoneStateListener extends PhoneStateListener {

    private Context mContext;

    public MyPhoneStateListener(Context context) {
        mContext = context;
    }

    @Override
    public void onServiceStateChanged(ServiceState serviceState) {
        super.onServiceStateChanged(serviceState);
    }

    @Override
    public void onCallStateChanged(int state, String incomingNumber) {

        switch (state) {
            case TelephonyManager.CALL_STATE_IDLE:         //电话挂断
                break;
            case TelephonyManager.CALL_STATE_RINGING:      //电话响铃
                break;
```

```
                    case TelephonyManager.CALL_STATE_OFFHOOK:    //来电接通或者去电
                        break;
                }
            }
        }
```

Android 可以监听到用户电话的拨打状态，从而做出相应的操作，可以监听到三个状态：无操作状态 CALL_STATE_IDLE、通话中状态 CALL_STATE_OFFHOOK、响铃中状态 CALL_STATE_RINGING。

9.3.4 挂断电话

使用系统服务提供的接口去挂断电话并不能保证成功，所以会有多种挂断方式同时使用，举例如下：

```
import android.content.Context;
import android.os.RemoteException;
import android.telephony.TelephonyManager;

import java.lang.reflect.InvocationTargetException;
import java.lang.reflect.Method;
import java.util.concurrent.Executor;
import java.util.concurrent.Executors;

public class HangUpTelephonyUtil {
    public static boolean endCall(Context context) {
        boolean callSuccess = false;
        ITelephony telephonyService = getTelephonyService(context);
        try {
            if (telephonyService != null) {
                callSuccess = telephonyService.endCall();
            }
        } catch (RemoteException e) {
            e.printStackTrace();
        } catch (Exception e){
            e.printStackTrace();
        }
        if (callSuccess == false) {
            Executor eS = Executors.newSingleThreadExecutor();
            eS.execute(new Runnable() {
                @Override
                public void run() {
                    disconnectCall();
                }
            });
            callSuccess = true;
        }
```

```
        return callSuccess;
    }

    //通过反射获取电话信息
    private static ITelephony getTelephonyService(Context context) {
        TelephonyManager telephonyManager = (TelephonyManager)
            context.getSystemService(Context.TELEPHONY_SERVICE);
        Class clazz;
        try {
            clazz = Class.forName(telephonyManager.getClass().getName());
            Method method = clazz.getDeclaredMethod("getITelephony");
            method.setAccessible(true);
            return (ITelephony) method.invoke(telephonyManager);
        } catch (ClassNotFoundException e) {
            e.printStackTrace();
        } catch (NoSuchMethodException e) {
            e.printStackTrace();
        } catch (IllegalArgumentException e) {
            e.printStackTrace();
        } catch (IllegalAccessException e) {
            e.printStackTrace();
        } catch (InvocationTargetException e) {
            e.printStackTrace();
        }
        return null;
    }

    //挂断电话
    private static boolean disconnectCall() {
        Runtime runtime = Runtime.getRuntime();
        try {
            runtime.exec("service call phone 5 \n");
        } catch (Exception exc) {
            exc.printStackTrace();
            return false;
        }
        return true;
    }

    //使用endCall挂断不了，再使用killCall反射调用再挂断一次
    public static boolean killCall(Context context) {
        try {
            //取得旧的电话管理者
            TelephonyManager telephonyManager = (TelephonyManager)
                context.getSystemService(Context.TELEPHONY_SERVICE);
```

```
            //取得getITelephony()方法
            Class classTelephony = Class.forName(telephonyManager.getClass().getName());
            Method methodGetITelephony = classTelephony.getDeclaredMethod("getITelephony");

            methodGetITelephony.setAccessible(true);

            Object telephonyInterface = methodGetITelephony.invoke(telephonyManager);

            //取得结束通话方法
            Class telephonyInterfaceClass = Class.forName(telephonyInterface.getClass().getName());
            Method methodEndCall = telephonyInterfaceClass.getDeclaredMethod("endCall");

            //调用endCall()
            methodEndCall.invoke(telephonyInterface);
        } catch (Exception ex) {
            //错误处理
            return false;
        }
        return true;
    }
}
```

ITelephony 接口在 layoutlib.jar 包中,需要导入 android sdk 目录\platforms\android-8\data\layoutlib.jar。

挂断电话需要权限,代码如下:

`<uses-permission android:name="android.permission.CALL_PHONE" />`

9.4 关键知识点解析

9.4.1 进程通信——AIDL 的使用

1. AIDL 的定义

Android 系统中的进程之间不能共享内存,因此需要提供一些机制在不同进程之间进行数据通信。AIDL(Android Interface Definition Language)是 Android 提供的一种进程间通信(IPC)机制。

Android 应用程序组件 Service 可以进行跨进程访问,Android 系统采用了远程过程调用(Remote Procedure Call,RPC)方式来实现。AIDL 可以使其他的应用程序跨进程访问本应用程序提供的服务。

通过这种机制,我们只需要写好 AIDL 接口文件,编译时系统会自动编译生成对应的 Binder 接口。

AIDL 支持以下 5 种数据类型:

- Java 的原生类型。

- String 和 CharSequence。
- List 和 Map，List 和 Map 对象的元素必须是 AIDL 支持的数据类型。
- AIDL 自动生成的接口。
- Parcelable 接口的类。

2. AIDL 的实现

（1）创建 AIDL。创建 AIDL 文件夹及实现 Parcelable 接口操作的实体类，然后编译项目系统会自动生成对应的 Binder Java 文件。

（2）服务端。建立一个服务类（Service 的子类），其他进程（应用）将远程调用的在其中创建上面生成的 Binder 对象实例，实现接口定义的方法在 onBind()中返回。在 AndroidManifest.xml 文件中配置 AIDL 服务，需要注意的是，标签中 android:name 的属性值就是客户端要引用该服务的 ID，也就是 Intent 类的参数值。

（3）客户端。将对应的 AIDL 类复制到工程中，实现 ServiceConnection 接口，然后绑定 AIDL 服务，并获得 AIDL 服务对象，最后调用 AIDL 服务的方法。

3. AIDL 实例

（1）创建 Parcelable 接口操作的实体类。创建 Parcelable 接口操作的实体类，具体代码如下：

```java
package com.jpy.myapplication;

import android.os.Parcel;
import android.os.Parcelable;

public class Boy implements Parcelable {
    private String mName;
    private int mAge;

    public Boy(String name,int age) {
        mName = name;
        mAge = age;
    }

    protected Boy(Parcel in) {
        mName = in.readString();
    }

    public int getmAge() {
        return mAge;
    }

    public void setmAge(int mAge) {
        this.mAge = mAge;
    }

    public static final Creator<Boy> CREATOR = new Creator<Boy>() {
```

```
        @Override
        public Boy createFromParcel(Parcel in) {
            return new Boy(in);
        }

        @Override
        public Boy[] newArray(int size) {
            return new Boy[size];
        }
    };

    @Override
    public int describeContents() {
        return 0;
    }

    @Override
    public void writeToParcel(Parcel dest, int flags) {
        dest.writeString(mName);
        dest.writeInt(mAge);
    }

    @Override
    public String toString() {
        return "Boy{" +
                "mName='" + mName + '\'' +
                "mAge='" + mAge + '\'' +
                '}';
    }
}
```

（2）新建 AIDL 文件夹，创建接口 AIDL 文件以及实体类对应的 AIDL 文件，其包名要与 Java 文件夹的包名一致，如图 9-14 所示。

图 9-14 新建 AIDL 文件夹

在 com.jpy.myapplication.aidl.Boy 中创建实体类的映射 AIDL 文件 BoyAidl.aidl，在其中声明映射的实体类名称与类型，然后创建对应的 AIDL 接口文件 myAidl.aidl，代码如下：

```
import com.jpy.myapplication.aidl.Boy.Boy;

interface myAidl {
    void addBoy(in Boy boy);
```

List<Boy> getBoyList();
}

注意，这个 BoyAidl.aidl 的包名要和实体类包名一致，除了基本数据类型，其他类型的参数都需要标上方向类型：in（输入）、out（输出）、inout（输入输出）。

在接口 AIDL 文件中定义两个要实现的接口：

- AddBoy：添加 boy。
- getBoyList：获取 boy 列表。

（3）Make Project，生成 Binder 的 Java 文件。点击 Build→Make Project，构建完成后系统会在 build/generated/source/aidl/用户的 flavor/下自动生成 myAidl.java 文件。AIDL 真正的强大之处就在这里。

1）服务端实现。建立一个 IMyService.aidl 文件，在其中实现了 AIDL 接口中定义的方法，代码如下：

```
public class IMyService extends Service {

    private ArrayList<Boy> mBoy;

    /**
     * 创建生成的本地Binder对象，实现AIDL制定的方法
     */
    private IBinder mIBinder = new IMyAidl.Stub() {

        @Override
        public void adBoy(Boy boy) throws RemoteException {
            mPersons.add(Boy);
        }

        @Override
        public List<Boy> getBoyList() throws RemoteException {
            return mBoy;
        }
    };

    @Nullable
    @Override
    public IBinder onBind(Intent intent) {
        mBoy = new ArrayList<>();
        return mIBinder;
    }
}
```

上面的代码中，onBind 是客户端与服务端绑定时的回调方法，返回 mIBinder 后客户端就可以通过它远程调用服务端的方法，即实现了通信。

服务需要在 Manifest 文件中声明，代码如下：

```
<service
    android:name=" com.jpy.myapplication.service. IMyService"
```

android:enabled="true"   android:exported="true"   android:process=":aidl"/>

2）客户端实现。实现 ServiceConnection 接口，在其中拿到 AIDL 类，代码如下。

```
private myAidl myaidl;

private ServiceConnection mConnection = new ServiceConnection() {
    @Override
    public void onServiceConnected(ComponentName name, IBinder service) {

        myaidl = myAidl .Stub.asInterface(service);
    }

    @Override
    public void onServiceDisconnected(ComponentName name) {
        myaidl = null;
    }
}
```

在 Activity 中创建一个服务连接对象，在其中调用 IMyAidl.Stub.asInterface()方法将 Binder 转换成 AIDL 类，在不同进程会返回一个代理。

3）绑定服务，代码如下。

```
Intent intent = new Intent(mContext, IMyService.class);
bindService(intent , mConnection, BIND_AUTO_CREATE);
```

注意：5.0 以上版本要求显式调用 Service，所以我们无法通过 action 或者 filter 的形式调用 Service。

4）实现 AIDL 类定义的接口，完成跨进程操作，代码如下。

```
public void addBoy() {
    Boy boy = new Boy("王强" , 15);
    mAidl. addBoy ( boy );
}
```

9.4.2 双卡双待手机如何获取来电

对于双卡双待手机，每张卡都对应一个 Service 和一个 PhoneStateListener，需要给每个服务注册自己的 PhoneStateListener，服务的名称还会有点变化，厂商可能会修改。

```
public ArrayList<String> getMultSimCardInfo() {
    //获取双卡的信息
    ArrayList<String> phoneServerList = new ArrayList<String>();
    for(int i = 1; i < 3; i++) {
        try {
            String phoneServiceName;
            if (MiuiUtils.isMiuiV6()) {
                phoneServiceName = "phone." + String.valueOf(i-1);
            } else {
                phoneServiceName = "phone" + String.valueOf(i);
            }
```

```
        //尝试获取服务
        IBinder iBinder = ServiceManager.getService(phoneServiceName);
        if(iBinder == null) continue;

        ITelephony iTelephony = ITelephony.Stub.asInterface(iBinder);
        if(iTelephony == null) continue;

        phoneServerList.add(phoneServiceName);
    } catch(Exception e) {
        e.printStackTrace();
    }
}
//这个是默认的
phoneServerList.add(Context.TELEPHONY_SERVICE);
return phoneServerList;
}
```

9.5　问题与讨论

1. 如何退出 Activity？如何安全退出调用多个 Activity 的 Application？
2. Service 是否在 main thread 中？Service 中能否进行耗时操作？
3. ContentProvider 是如何实现数据共享的？
4. 如何启动其他应用的 Service？

项目 10　蓝牙打印机

本项目设计一款集休闲娱乐办公于一体的打印机软件的应用,既可以随手打印花式图片,也可以打印办公便条。该应用具有打印文字和图片、打印网页、涂鸦和分享打印等功能。

项目需求描述如下:

(1)用户注册和登录模块。

(2)文字打印模块,功能包括要打印文字的大小和对齐方式。

(3)可以打印手机分享出来的文字和图片。

(4)网页打印模块,功能包括网址的转入、前进和后退,网页显示字体的放大和缩小,打印预览,当前页打印和全部打印。

(5)涂鸦模块,功能包括设置画笔粗细、操作撤销和擦除。

(6)打印图片模块,功能包括拍照打印和本地相册打印。

(7)蓝牙搜索模块,功能包括蓝牙搜索、选择和匹配。

(8)可以设置软件语言。

(1)学习蓝牙通信。

(2)学习蓝牙搜索匹配。

10.1　总体设计

10.1.1　总体分析

该应用通过蓝牙模块实现手机与打印机的交互通信,控制打印机进行文字、图片、网页和涂鸦的打印。

10.1.2　功能模块框图

根据总体分析结果可以总结出功能模块框图,如图 10-1 所示。

总体模块的作用主要是生成应用程序的主类,控制应用程序的生命周期;基础架构模块主要提供程序架构、所有 Activity 公用的父类、所有 Activity 公用的方法,包括自定义风格对话框、自定义提示框等功能；数据管理与控制模块主要提供数据获取、数据解析、数据读写、数据组织和数据缓存功能；用户界面模块包括打印文字、图片、涂鸦、网页、操作提示等功能；网络通信模块主要负责获取网页信息。

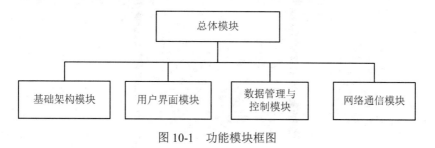

图 10-1　功能模块框图

10.1.3　系统流程图

根据总体分析结果及功能模块框图梳理出系统启动的主要流程，如图 10-2 所示。

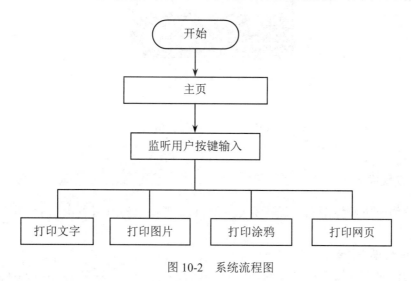

图 10-2　系统流程图

10.1.4　界面设计

本应用是操作打印机的应用，主要实现打印文字、图片、涂鸦和网页。

根据程序功能需求可以规划出软件的主要界面，如下：

（1）主页：显示文字、涂鸦、网页和设置。

（2）文字：点击主页中的"文字"按钮进入文字打印页面。

（3）涂鸦：点击主页中的"涂鸦"按钮进入涂鸦打印页面。

（4）网页：点击主页中的"网页"按钮进入网页打印页面。

系统主界面如图 10-3 所示。从图 10-3 中可以很直观地看到，主界面包含文字、涂鸦、网页和设置 4 个功能。

图 10-3　系统主界面

10.2　详细设计

10.2.1　模块描述

在系统总体分析及界面布局设计完成后，主要工作就转入对各个功能模块的详细设计阶段。

1. 基础架构模块详细设计

基础架构模块主要提供程序架构、所有 Activity 公用的父类、所有 Activity 公用的方法，包括自定义风格对话框、自定义提示框等功能。

基础架构模块功能如图 10-4 所示。

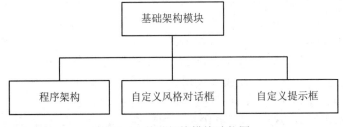

图 10-4　基础架构模块功能图

2. 用户界面模块详细设计

用户界面模块的主要任务是显示打印内容设置页面和实现与用户的交互，即当用户点击按键或者屏幕的时候，监听器会去调用相应的处理办法或其他相应的处理模块。

本模块包括文字打印、涂鸦打印、网页打印等功能。

用户界面模块功能如图 10-5 所示。

图 10-5　用户界面模块功能图

3. 数据管理与控制模块详细设计

数据管理与控制模块主要提供数据获取、数据解析、数据读写、数据组织和**数据缓存**功能。

数据管理与控制模块和用户界面模块可以调用基础架构模块的一些通用方法。**数据管理与控制模块为用户界面模块提供数据，同时可以接收并保存用户界面模块产生的数据。**

数据管理与控制模块功能如图 10-6 所示。

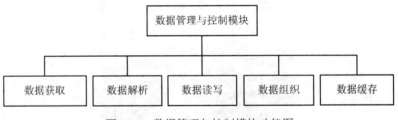

图 10-6　数据管理与控制模块功能图

4. 网络通信模块详细设计

网络通信模块根据用户界面的需求调用访问服务器，接收服务器返回的数据并解析，同时显示到用户界面。

本模块包括发送网络请求、接收网络应答、网络数据解析等功能。

网络通信模块功能如图 10-7 所示。

图 10-7　网络通信模块功能图

10.2.2 系统包及其资源规划

1. 文件结构

在系统各个模块的实现方式和流程设计完成后，就可以对系统主要的包和资源进行规划，划分的原则主要是保持各个包相互独立，耦合度尽量低。

根据系统功能设计，本系统封装一个基础的 Activity 类，加载各个子页的通用控件，并提供一些基础的实现方法，例如设置进度条、标题等常用的方法，程序中的 Activity 都可继承此基类，继承后就可直接使用基类中封装的基础方法。

系统使用 6 个 Activity、一个 Activity 用于主页，一个 Activity 用于文字打印，一个 Activity 用于涂鸦打印，一个 Activity 用于网页打印，一个 Activity 用于分享打印，一个 Activity 用于设置。包及其资源结构如图 10-8 所示。

图 10-8　包及其资源结构

2. 命名空间

本示例设置了多个命名空间，分别用来保存用户界面、后台服务的源代码文件，具体说明见表 10-1。

表 10-1 命名空间

命名空间	说明
cn.chinaunicom.btprintchen.Bean	存放与打印信息相关的源代码文件
cn.chinaunicom.btprintchen.Adapter	存放与页面适配器相关的源代码文件
cn.chinaunicom.btprintchen.PopMenu	存放与自定义控件相关的源代码文件
cn.chinaunicom.btprintchen.utils	存放与工具类相关的源代码文件
cn.chinaunicom.btprintchen.ui	存放与页面视图相关的源代码文件

3. 源代码文件

源代码文件及说明见表 10-2。

表 10-2 源代码文件

包名称	文件名	说明
cn.chinaunicom.btprintchen.Bean	BeanTxt.java	文字打印信息类
cn.chinaunicom.btprintchen.Adapter	Bt_listAdapter.java	蓝牙列表适配器类
	TxtAdapter.java	文字打印信息适配器类
	AppsAdapter.java	主页 Grid 适配器类
cn.chinaunicom.btprintchen.PopMenu	PopMenu.java	自定义弹出菜单类
	SimpleAppsGridView.java	自定义 Grid 类
cn.chinaunicom.btprintchen.utils	StringUtilsZhy.java	字符串工具类
	ImageUtils.java	图片工具类
	Tools.java	自定义工具栏高度类
cn.chinaunicom.btprintchen.ui	MainActivity.java	主页 Activity
	SplashActivity.java	欢迎页 Activity
	TxtActivity.java	文字打印页 Activity
	PrintSetActivity.java	设置页 Activity
	UrlActivity	网页打印页 Activity
	ImageActivity	图片打印页 Activity
	DoodleActivity.java	涂鸦打印页 Activity

4. 资源文件

Android 的资源文件保存在/res 的子目录中。

- /res/drawable 目录：保存的是图像文件。
- /res/layout 目录：保存的是布局文件。
- /res/values 目录：保存的是用来定义字符串和颜色的文件。

资源文件及说明见表 10-3。

表 10-3　资源文件

资源目录	文件	说明
drawable	splash.png	欢迎页背景
	search_nor.png	搜索图标
	back_rtr.png	右箭头图标
	ic_launcher.png	程序图标文件
	bg.jpg	主界面背景图
	icon_url.png	网页按钮
	icon_txt.png	文字按钮
	icon_set.png	设置按钮
	icon_pen_sel.png	画笔按钮
	icon_eraser_sel.png	橡皮擦按钮
	ic_bluetooth.png	蓝牙图标
	icon_clear_sel.png	删除按钮
	icon_paint.png	涂鸦按钮
layout	activity_main.xml	主界面的布局
	activity_txt.xml	文字打印页面的布局
	activity_url.xml	网页打印页面的布局
	activity_doodle	涂鸦打印页面的布局
	activity_image	图片打印页面的布局
	activity_image_preview.xml	图片预览页面的布局
	activity_welcome.xml	欢迎页的布局
values	dimens.xml	保存尺寸的 XML 文件
	attrs.xml	保存样式的 XML 文件
	colors.xml	保存颜色的 XML 文件
	strings.xml	保存字符串的 XML 文件
	styles.xml	保存样式的 XML 文件

10.2.3　主要方法流程设计

蓝牙通信流程图如图 10-9 所示。

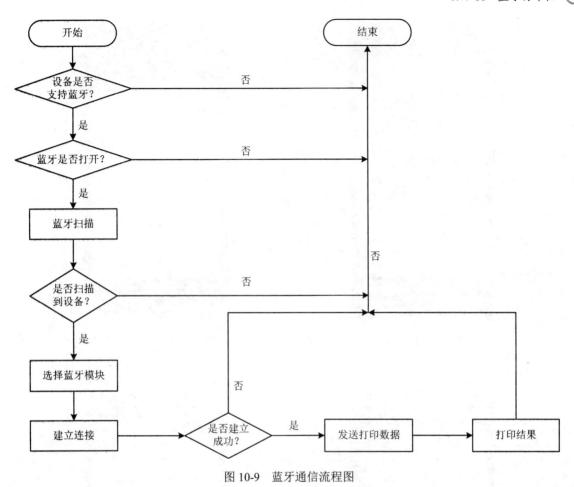

图 10-9 蓝牙通信流程图

10.3 代码实现

10.3.1 显示界面布局

1. 系统主界面

系统主界面是进入系统后显示的界面，该界面包括一个 Grid 控件，如图 10-10 所示。

2. 文本打印界面

文本打印界面可以对文本进行字体大小设置和打印，该界面包括三个 Button、一个 RadioGroup 和一个 EditText，如图 10-11 所示。

3. 涂鸦打印界面

涂鸦打印界面可以对涂鸦进行笔画粗细设置、擦除、撤销和打印，该界面包括 4 个 Button 和一个 View，如图 10-12 所示。

4. 网页打印界面

网页打印界面可以对打印网页进行网址转入、前进和后退操作，网页文字字体大小设置和网页打印设置，该界面包括 6 个 Button、一个 EditText 和一个 WebView，如图 10-13 所示。

图 10-10　系统主界面　　　　　　　图 10-11　文本打印界面

图 10-12　涂鸦打印界面　　　　　　图 10-13　网页打印界面

5. 图片打印界面

图片打印界面可以对图片进行预览和打印，该界面包括一个 Button 和一个 Imageview，如图 10-14 所示。

图 10-14　图片打印界面

10.3.2　控件设计实现

自定义弹出菜单，代码如下：

```
public abstract class PopMenu {
    /**
     * 上下文
     */
    private Context mContext;
    /**
     * 菜单项
     */
    private ArrayList<Item> mItemList;
    /**
     * 列表适配器
     */
    private ArrayAdapter<Item> mAdapter;
    /**
     * 菜单选择监听
     */
    private OnItemSelectedListener mListener;
    /**
     * 列表
     */
    private ListView mListView;
    /**
     * 弹出窗口
     */
```

```java
        private PopupWindow mPopupWindow;

        public PopMenu(Context context) {
            mContext = context;
            mItemList = new ArrayList<Item>(2);
            View view = onCreateView(context);
            view.setFocusableInTouchMode(true);
            mAdapter = onCreateAdapter(context, mItemList);
            mListView = findListView(view);
            mListView.setAdapter(mAdapter);
            mListView.setOnItemClickListener(new AdapterView.OnItemClickListener() {
                @Override
                public void onItemClick(AdapterView<?> parent, View view, int position, long id) {
                    Item item = mAdapter.getItem(position);
                    if (mListener != null) {
                        mListener.selected(view, item, position);
                    }
                    mPopupWindow.dismiss();
                }
            });
            view.setOnKeyListener(new View.OnKeyListener() {
                @Override
                public boolean onKey(View v, int keyCode, KeyEvent event) {
                    if (keyCode == KeyEvent.KEYCODE_MENU && mPopupWindow.isShowing()) {
                        mPopupWindow.dismiss();
                        return true;
                    }
                    return false;
                }
            });
            mPopupWindow = new PopupWindow(view, ViewGroup.LayoutParams.WRAP_CONTENT,
                    ViewGroup.LayoutParams.WRAP_CONTENT, true);
            mPopupWindow.setBackgroundDrawable(new ColorDrawable(0x00000000));
        }

        /**
         * 菜单的界面视图
         *
         * @param context
         * @return
         */
        protected abstract View onCreateView(Context context);

        /**
         * 菜单界面视图中的列表
         *
```

```
 * @param view
 * @return
 */
protected abstract ListView findListView(View view);

/**
 * 菜单列表中的适配器
 *
 * @param context
 * @param itemList表示所有菜单项
 * @return
 */
protected abstract ArrayAdapter<Item> onCreateAdapter(Context context, ArrayList<Item> itemList);

/**
 * 添加菜单项
 *
 * @param text表示菜单项文字内容
 * @param id表示菜单项的ID
 */
public void addItem(String text, int id) {
    mItemList.add(new Item(text, id));
    mAdapter.notifyDataSetChanged();
}

/**
 * 添加菜单项
 *
 * @param resId表示菜单项文字内容的资源ID
 * @param id表示菜单项的ID
 */
public void addItem(int resId, int id) {
    addItem(mContext.getString(resId), id);
}

/**
 * 作为指定View的下拉控制显示
 *
 * @param parent表示所指定的View
 */
public void showAsDropDown(View parent) {
    mPopupWindow.showAsDropDown(parent);
}

/**
 * 隐藏菜单
```

```java
         */
        public void dismiss() {
            mPopupWindow.dismiss();
        }

        /**
         * 设置菜单选择监听
         *
         * @param listener表示监听器
         */
        public void setOnItemSelectedListener(OnItemSelectedListener listener) {
            mListener = listener;
        }

        /**
         * 当前菜单是否正在显示
         *
         * @return
         */
        public boolean isShowing() {
            return mPopupWindow.isShowing();
        }

        /**
         * 菜单项
         */
        public static class Item {
            public String text;
            public int id;

            public Item(String text, int id) {
                this.text = text;
                this.id = id;
            }

            @Override
            public String toString() {
                return text;
            }
        }

        /**
         * 菜单项选择监听接口
         */
        public static interface OnItemSelectedListener {
            /**
```

```
     * 菜单被选择时的回调接口
     *
     * @param view表示被选择的内容的View
     * @param item表示被选择的菜单项
     * @param position表示被选择的位置
     */
    public void selected(View view, Item item, int position);
    }
}
```

10.3.3 获取图片分享

应用需要接收分享来的图片进行打印，具体做法：在 ImageActivity 的清单文件里面设置 action、category 和 data。

```
<intent-filter>
    <action android:name="android.intent.action.SEND"/>
    <category android:name="android.intent.category.DEFAULT"/>
    <data android:mimeType="image/*"/>
</intent-filter>
```

data 表示接收的文件类型，如果是文本类型则不会接收。这样进行标注后再分享内容时系统就会识别程序并展示给用户选择了。

获取分享的内容和处理接收到的内容，代码如下：

```
private void handleImage() {
    Intent intent = getIntent();
    String action = intent.getAction();
    String type = intent.getType();
    String clipdata="";
    if(intent.getClipData()!=null &&intent.getClipData().getItemCount()>0 ) {
        clipdata = intent.getClipData().getItemAt(0).toString();
        if (clipdata.indexOf("http")>0)
        {
            int index = clipdata.indexOf("http://") >= 0 ? clipdata.indexOf("http://") :
                clipdata.indexOf("https://");
            int index2 = clipdata.indexOf("}");
            clipdata=clipdata.substring(index,index2);
            startActivity(new Intent(Main2Activity.this, UrlActivity.class).putExtra("data", clipdata));

            finish();
            return;
        }
    }
    if (action != null)
        switch (action) {
            case Intent.ACTION_SEND: {
                if (type.startsWith("image/")) {
```

```java
            Uri uri = intent.getParcelableExtra(Intent.EXTRA_STREAM);
            if (uri != null) {
                Log.e("url", uri.toString());
                startActivity(new Intent(Main2Activity.this, ImageEditActivity.class).putExtra
                        ("imagePath", getRealFilePath(Main2Activity.this, uri)));
                finish();
            }

        } else if (type.equals("text/plain")) {
            final String txt = intent.getStringExtra(Intent.EXTRA_TEXT);
            if (txt != null) {
                Log.e("url", txt.toString());
                if (txt.indexOf("http://") >= 0 || txt.indexOf("https://") >= 0) {
                    AlertDialog.Builder builder = new AlertDialog.Builder(Main2Activity.this);
                    builder.setMessage(getString(R.string.txt_t4));
                    builder.setTitle(getString(R.string.bluetooth_t2));
                    builder.setPositiveButton(getString(R.string.txt_t5),
                        new DialogInterface.OnClickListener() {
                          @Override
                          public void onClick(DialogInterface dialog, int which) {
                              int index = txt.indexOf("http://") >= 0 ? txt.indexOf("http://") :
                                txt.indexOf("https://");
                              startActivity(new Intent(Main2Activity.this, UrlActivity.class).
                                      putExtra("data", txt.substring(index)));
                              dialog.dismiss();
                              finish();
                          }
                    });
                    builder.setNegativeButton(getString(R.string.txt_t6),
                        new DialogInterface.OnClickListener() {
                          @Override
                          public void onClick(DialogInterface dialog, int which) {
                              startActivity(new Intent(Main2Activity.this, TxtActivity.class).
                                      putExtra("data", txt));
                              dialog.dismiss();
                              finish();
                          }
                    });
                    builder.create().show();
                } else {
                    startActivity(new Intent(Main2Activity.this, TxtActivity.class).
                        putExtra("data", txt));
                    finish();
                }
            }
        }
```

```
        }
      }
    }
}
```

通过 getIntent 方法获取到包含分享内容的 Intent，然后就可以获取里面的内容。如果分享的图片是在 SD 卡中，程序需要添加读取 SD 卡的权限，否则会显示 permission denied。

```
<uses-permission android:name="android.permission.WRITE_EXTERNAL_STORAGE"/>
```

以上说的都是分享一条文本或图片，如果要分享多个，方法是一样的。只需在分享时用 ArratList 进行封装即可，对于接收者来说，需要把 action 改成如下代码：

```
<action android:name="android.intent.action.SEND_MULTIPLEND"/>
```

以上代码表示接收多个内容。在代码里面从 Intent 中获取内容时用 getParcelableArrayListExtra 而不是 getParcelableExtra。

10.3.4 蓝牙设备和设置可见时间

为蓝牙设备设置可见时间，代码如下：

```
private void setBluetoothShowTime() {
        BluetoothAdapter adapter = BluetoothAdapter.getDefaultAdapter();
        if (!adapter.isEnabled()) {
            adapter.enable();
        }
        Intent enable = new Intent(BluetoothAdapter.ACTION_REQUEST_DISCOVERABLE);
        enable.putExtra(BluetoothAdapter.EXTRA_DISCOVERABLE_DURATION, 60);
        startActivity(enable);
}
```

10.3.5 搜索蓝牙设备

要想与蓝牙模块进行通信，首先得搜到该设备：
```
bluetoothAdapter.startDiscovery();

private class BluetoothReceiver extends BroadcastReceiver {
    @Override
    public void onReceive(Context context, Intent intent) {
        String action = intent.getAction();
        if (BluetoothDevice.ACTION_FOUND.equals(action)) {
            BluetoothDevice device = intent.getParcelableExtra(BluetoothDevice.EXTRA_DEVICE);
            if (isLock(device)) {
                devices.add(device.getName());
            }
            deviceList.add(device);
        }
        showDevices();
    }
}
```

startDiscovery()是一个异步方法，它会对其他蓝牙设备进行搜索，搜索过程其实是在 System

Service 中进行的,我们可以通过 cancelDiscovery()方法来停止这个搜索。

在 Android 中使用广播需要注册,代码如下:

```
///广播注册
IntentFilter filter = new IntentFilter(BluetoothDevice.ACTION_FOUND);
receiver = new BluetoothReceiver();
registerReceiver(receiver, filter);
    //结束撤销广播服务监听
    @Override
    protected void onDestroy() {
        unregisterReceiver(receiver);
        super.onDestroy();
    }
```

10.3.6 连接蓝牙设备

搜索到该设备后,就要对该设备进行连接,代码如下:

```
public int openPrinter(String Deviveid, String Pwd) {
    int res = 0;
    this.mBtAdapter = BluetoothAdapter.getDefaultAdapter();
    if (this.mBtAdapter == null) {
        return -1;
    } else {
        this.isPrinting = false;
        boolean isopen = true;
        if (!this.mBtAdapter.isEnabled()) {
            this.mBtAdapter.enable();
            isopen = false;
        }

        if (Deviveid.length() == 0) {
            res = 2003;
            return res;
        } else {
            this.device = this.mBtAdapter.getRemoteDevice(Deviveid);
            if (this.device.getBondState() != 12) {
                try {
                    this.autoBond(this.device.getClass(), this.device, Pwd);
                    this.createBond(this.device.getClass(), this.device);
                } catch (Exception var8) {
                    System.out.println("配对不成功");
                    res = 2003;
                }
            }

            try {
                this.btSocket = this.device.createRfcommSocketToServiceRecord(MY_UUID);
```

```
                this.btSocket.connect();
                this.outStream = this.btSocket.getOutputStream();
                this.inStream = this.btSocket.getInputStream();
            } catch (IOException var7) {
                var7.printStackTrace();
                res = 2002;
            }

            if (res != 0 && !isopen) {
                try {
                    Thread.sleep(2000L);
                } catch (InterruptedException var6) {
                    var6.printStackTrace();
                }
            }

            return res;
        }
    }
}
```

连接设备之前需要 UUID。UUID（Universally Unique Identifier）是一个 128 位的字符串 ID，用于进行唯一标识，通常使用以下 UUID 进行连接。

```
private static final UUID MY_UUID = UUID.fromString("00001101-0000-1000-8000-00805f9b34fb");
```

10.3.7　蓝牙通信

如果连接没有问题，我们就可以和蓝牙模块进行通信了，代码如下：

```
if (isConnect) {
    try {
        OutputStream outStream = socket.getOutputStream();
        outStream.write(getHexBytes(message));
    } catch (IOException e) {
        setState(WRITE_FAILED);
        Log.e("TAG", e.toString());
    }
    try {
        InputStream inputStream = socket.getInputStream();
        int data;
        while (true) {
            try {
                data = inputStream.read();
                Message msg = handler.obtainMessage();
                msg.what = DATA;
                msg.arg1 = data;
                handler.sendMessage(msg);
            } catch (IOException e) {
                setState(READ_FAILED);
```

```
                    Log.e("TAG", e.toString());
                    break;
                }
            }
        } catch (IOException e) {
            setState(WRITE_FAILED);
            Log.e("TAG", e.toString());
        }
    }

        if (socket != null) {
            try {
                socket.close();
            } catch (IOException e) {
                Log.e("TAG", e.toString());
            }
        }
```

蓝牙通信所需权限如下：

`<uses-permission android:name="android.permission.BLUETOOTH"/>`
`<uses-permission android:name="android.permission.BLUETOOTH_ADMIN"/>`

10.4 关键知识点解析

10.4.1 静默开启蓝牙

开启蓝牙对话框如图 10-15 所示。

图 10-15　开启蓝牙对话框

如果不想让用户看到这个对话框，则可以静默开启蓝牙。
在 AndroidManifest 文件中添加需要的权限：

`<uses-permission android:name="android.permission.BLUETOOTH" />`
`<uses-permission android:name="android.permission.BLUETOOTH_ADMIN" />`
`<uses-permission android:name="android.permission.ACCESS_FINE_LOCATION" />`
`<uses-permission android:name="android.permission.ACCESS_COARSE_LOCATION" />`
`<uses-feature`
`   android:name="android.hardware.bluetooth_le"`
`   android:required="true" />`

由于蓝牙所需要的权限包含危险权限，所以我们需要在 Java 代码中进行动态授权处理：
private static final int REQUEST_BLUETOOTH_PERMISSION=1;

```
private void checkBluetoothPermission(){
    //判断系统版本
    if (Build.VERSION.SDK_INT >= 23) {
        //检测当前App是否拥有某个权限
        int checkBlueToothPermission = ContextCompat.checkSelfPermission(this,
            Manifest.permission.ACCESS_COARSE_LOCATION);
        //判断这个权限是否已经授权过
        if(checkBlueToothPermission != PackageManager.PERMISSION_GRANTED){
            if(ActivityCompat.shouldShowRequestPermissionRationale(this,
                Manifest.permission.ACCESS_COARSE_LOCATION))
            ActivityCompat.requestPermissions(this ,new String[]
                {Manifest.permission.ACCESS_COARSE_LOCATION},REQUEST_BLUETOOTH_PERMISSION);
            return;
        }
    }
}
```

10.4.2　蓝牙自动配对

正常蓝牙连接时需要提示用户输入配对密码，如果预置密码如何进行自动连接呢？

在自动匹配的时候可以通过反射调用 BluetoothDevice 的 setPin、createBond、cancelPairingUserInput 实现设置密钥、配对请求创建、取消密钥信息输入等，从而实现蓝牙的自动连接。

在 Android 源码里面有一个自动配对的方法，也就是把 Pin 值自动设为 0000，代码如下：

```
synchronized boolean attemptAutoPair(String address) {
        if (!mBondState.hasAutoPairingFailed(address) &&
                !mBondState.isAutoPairingBlacklisted(address)) {
            mBondState.attempt(address);
            setPin(address, BluetoothDevice.convertPinToBytes("0000"));
            return true;
        }
        return false;
}
```

从该方法可以看出，实现自动设置是在底层回调到 Java 层的 onRequestPinCode 方法时被调用的，首先判断正常输入的密码，然后再对 0000 密码进行自动配对。

由于蓝牙设备配对的方法都是隐藏的，所以我们需要通过反射调用被隐藏的方法，代码如下：

```
/**
 * 与设备配对参考源码：platform/packages/apps/Settings.git
 * /Settings/src/com/android/settings/bluetooth/CachedBluetoothDevice.java
 */
static public boolean createBond(Class btClass, BluetoothDevice btDevice) throws Exception
```

```java
{
    Method createBondMethod = btClass.getMethod("createBond");
    Boolean returnValue = (Boolean) createBondMethod.invoke(btDevice);
    return returnValue.booleanValue();
}

/**
 * 与设备解除配对参考源码：platform/packages/apps/Settings.git
 * /Settings/src/com/android/settings/bluetooth/CachedBluetoothDevice.java
 */
static public boolean removeBond(Class<?> btClass, BluetoothDevice btDevice) throws Exception
{
    Method removeBondMethod = btClass.getMethod("removeBond");
    Boolean returnValue = (Boolean) removeBondMethod.invoke(btDevice);
    return returnValue.booleanValue();
}

static public boolean setPin(Class<? extends BluetoothDevice> btClass, BluetoothDevice btDevice,
        String str) throws Exception
{
    try
    {
        Method removeBondMethod = btClass.getDeclaredMethod("setPin", new Class[]
                {byte[].class});
        Boolean returnValue = (Boolean) removeBondMethod.invoke(btDevice, new Object[]
                {str.getBytes()});
    }
    catch (SecurityException e)
    {
        //throw new RuntimeException(e.getMessage());
        e.printStackTrace();
    }
    catch (IllegalArgumentException e)
    {
        //throw new RuntimeException(e.getMessage());
        e.printStackTrace();
    }
    catch (Exception e)
    {
        e.printStackTrace();
    }
    return true;

}

//取消用户输入
```

```
static public boolean cancelPairingUserInput(Class<?> btClass,
        BluetoothDevice device)    throws Exception
{
    Method createBondMethod = btClass.getMethod("cancelPairingUserInput");
    Boolean returnValue = (Boolean) createBondMethod.invoke(device);
    return returnValue.booleanValue();
}

//取消配对
static public boolean cancelBondProcess(Class<?> btClass, BluetoothDevice device) throws Exception
{
    Method createBondMethod = btClass.getMethod("cancelBondProcess");
    Boolean returnValue = (Boolean) createBondMethod.invoke(device);
    return returnValue.booleanValue();
}

//确认配对
static public void setPairingConfirmation(Class<?> btClass,BluetoothDevice device,boolean isConfirm)throws Exception
{
    Method setPairingConfirmation = btClass.getDeclaredMethod("setPairingConfirmation",boolean.class);
    setPairingConfirmation.invoke(device,isConfirm);
}
```

10.5　问题与讨论

1. 如何解决已连接的蓝牙无法断开的问题？
2. 蓝牙连接请求是否能在子线程中执行？如在子线程中执行应注意什么？
3. 如何进行 BLE（蓝牙低能耗）设备 Service 缓存刷新？
4. 多次扫描蓝牙，在部分机器上会扫描失败，如何避免这个问题？

项目 11　基于 Socket 的 Bmop 即时通信

本项目设计一款基于 Socket 通信技术的聊天小应用，可以实现好友的 IM 即时聊天功能。该应用具有添加好友和好友聊天功能。

项目需求描述如下：
（1）用户注册和登录模块。
（2）会话列表模块，显示最近好友会话列表，功能包括删除会话记录和显示未读消息。
（3）聊天模块，支持发送文字、图片、语音和位置。
（4）联系人模块，功能包括好友列表、添加好友和删除好友。
（5）设置模块，功能包括设置头像、昵称、性别及退出登录。

（1）学习 Socket 通信。
（2）学习消息队列。

11.1　总体设计

11.1.1　总体分析

该应用通过 Socket 实现好友的交互通信，用户可以和好友进行聊天，聊天支持发送文字、图片、语音和位置，可以添加好友、删除好友和编辑个人信息。

11.1.2　功能模块框图

根据总体分析结果可以总结出功能模块框图，如图 11-1 所示。

图 11-1　功能模块框图

总体模块的作用主要是生成应用程序的主类，控制应用程序的生命周期；基础架构模块主要提供程序架构、所有 Activity 公用的父类、所有 Activity 公用的方法，包括自定义风格对话框、自定义提示框等功能；数据管理与控制模块主要提供数据获取、数据解析、数据读写、数据组织和数据缓存功能；用户界面模块包括会话列表、联系人和设置等功能；网络通信模块主要负责同服务端进行数据交换，以实现聊天功能。

11.1.3 系统流程图

根据总体分析结果及功能模块框图梳理出系统启动的主要流程，如图 11-2 所示。

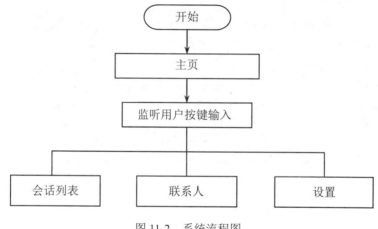

图 11-2　系统流程图

11.1.4 界面设计

本应用是 IM 即时通信应用，主要实现好友聊天。
根据程序功能需求可以规划出软件的主要界面，如下：
（1）会话：显示最近会话列表、显示未读消息数。
（2）联系人：好友列表、添加及删除好友。
（3）设置：编辑个人信息、退出登录。
系统主界面如图 11-3 所示。

图 11-3　系统主界面

从图 11-3 中可以很直观地看到，主页界面包含会话、联系人和设置三个功能。

11.2 详细设计

11.2.1 模块描述

在系统总体分析及界面布局设计完成后，主要工作就转入对各个功能模块的详细设计阶段。

1. 基础架构模块详细设计

基础架构模块主要提供程序架构、所有 Activity 公用的父类、所有 Activity 公用的方法，包括自定义风格对话框、自定义提示框等功能。

基础架构模块功能如图 11-4 所示。

图 11-4　基础架构模块功能图

2. 用户界面模块详细设计

用户界面模块的主要任务是显示会话、联系人、设置页及实现与用户的交互，即当用户点击按键或者屏幕的时候，监听器会去调用相应的处理办法或其他相应的处理模块。

本模块包括会话、联系人和设置等功能。

用户界面模块功能如图 11-5 所示。

图 11-5　用户界面模块功能图

3. 数据管理与控制模块详细设计

数据管理与控制模块主要提供数据获取、数据解析、数据读写、数据组织和数据缓存功能。

数据管理与控制模块和用户界面模块可以调用基础架构模块的一些通用方法。数据管理与控制模块为用户界面模块提供数据，同时可以接收并保存用户界面模块产生的数据。

数据管理与控制模块功能如图 11-6 所示。

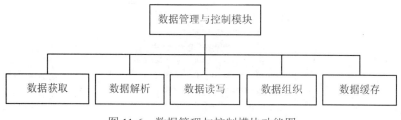

图 11-6　数据管理与控制模块功能图

4. 网络通信模块详细设计

网络通信模块根据用户界面的需求调用访问服务器，接收服务器返回的数据并解析，同时显示到用户界面。

本模块包括发送网络请求、接收网络应答、网络数据解析等功能。

网络通信模块功能如图 11-7 所示。

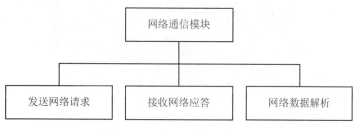

图 11-7　网络通信模块功能图

11.2.2　系统包及其资源规划

1. 文件结构

在系统各个模块的实现方式和流程设计完成后，就可以对系统主要的包和资源进行规划，划分的原则主要是保持各个包相互独立，耦合度尽量低。

根据系统功能设计，本系统封装一个基础的 Activity 类，加载各个子页的通用控件，并提供一些基础的实现方法，例如设置进度条、标题等常用的方法，程序中的 Activity 都可继承此基类，继承后就可直接使用基类中封装的基础方法。

系统使用 5 个 Activity 和 3 个 Fragment，一个 Activity 用于注册，一个 Activity 用于登录，一个 Activity 用于主页，一个 Activity 用于对话，一个 Activity 用于个人信息编辑，一个 Fragment 用于最近聊天记录，一个 Fragment 用于联系人，一个 Fragment 用于设置。包及其资源结构如图 11-8 所示。

2. 命名空间

本示例设置了多个命名空间，分别用来保存用户界面、后台服务的源代码文件，具体说明见表 11-1。

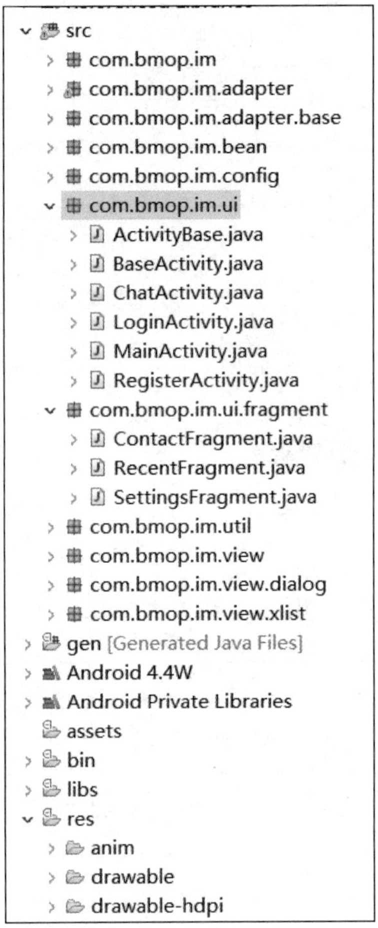

```
src
  com.bmop.im
  com.bmop.im.adapter
  com.bmop.im.adapter.base
  com.bmop.im.bean
  com.bmop.im.config
  com.bmop.im.ui
    ActivityBase.java
    BaseActivity.java
    ChatActivity.java
    LoginActivity.java
    MainActivity.java
    RegisterActivity.java
  com.bmop.im.ui.fragment
    ContactFragment.java
    RecentFragment.java
    SettingsFragment.java
  com.bmop.im.util
  com.bmop.im.view
  com.bmop.im.view.dialog
  com.bmop.im.view.xlist
gen [Generated Java Files]
Android 4.4W
Android Private Libraries
assets
bin
libs
res
  anim
  drawable
  drawable-hdpi
```

图 11-8　包及其资源结构

表 11-1　命名空间

命名空间	说明
com.bmop.im	存放与全局相关的源代码文件
com.bmop.im.adapter	存放与页面适配器相关的源代码文件
com.bmop.im.bean	存放与数据对象相关的源代码文件
com.bmop.im.config	存放与配置相关的源代码文件
com.bmop.im.ui	存放与页面视图相关的源代码文件
com.bmop.im.ui.fragment	存放与页面中的 Fragment 相关的源代码文件
com.bmop.im.util	存放与工具相关的源代码文件
com.bmop.im.view	存放与自定义 View 相关的源代码文件

3. 源代码文件

源代码文件及说明见表 11-2。

表 11-2 源代码文件

包名称	文件名	说明
com.bmop.im	CustomApplcation.java	全局 Applcation 类
com.bmop.im.adapter	MessageChatAdapter.java	聊天适配器类
	MessageRecentAdapter.java	会话适配器类
	UserFriendAdapter.java	好友列表适配器类
com.bmop.im.bean	User.java	个人信息类
	FaceText.java	表情类
com.bmop.im.config	BmopConstants.java	常量配置类
com.bmop.im.util	FaceTextUtils.java	表情工具类
	ImageUtils.java	图片工具类
	TimeUtil.java	时间工具类
cn.chinaunicom.btprintchen.ui	MainActivity.java	主页 Activity
	SplashActivity.java	欢迎页 Activity
	BaseActivity.java	Activity 基类
	ChatActivity.java	聊天页 Activity
com.bmop.im.ui.fragment	ContactFragment.java	联系人页 Fragment
	RecentFragment.java	会话页 Fragment
	SettingsFragment.java	设置页 Fragment
com.bmop.im.view	ClearEditText.java	带清除功能的输入框
	CustomGridView.java	自定义 GridView
	EmoticonsEditText.java	带表情输入功能的输入框
	DialogTips.java	自定义提示对话框
	MyLetterView.java	通讯录右侧快速滚动栏

4. 资源文件

Android 的资源文件保存在 /res 的子目录中。

- /res/drawable 目录：保存的是图像文件。
- /res/layout 目录：保存的是布局文件。
- /res/values 目录：保存的是用来定义字符串和颜色的文件。

资源文件及说明见表 11-3。

表 11-3 资源文件

资源目录	文件	说明
drawable	splash.png	欢迎页背景
	chat_fail_resend_normal.png	消息发送失败图标
	chat_keyboard_normal.png	键盘按钮
	ic_launcher.png	程序图标文件
	bg.jpg	主界面背景图
	chat_voice_normal.png	语音按钮
	default_head.png	默认头像
	icon_near.png	位置图标
	list_newmessage2.9.png	消息背景
	msg_tips.png	新信息提示
	ic_bluetooth.png	蓝牙图标
	new_friends_icon.png	添加好友按钮
	refresh_black.png	刷新
layout	activity_main.xml	主界面页面的布局
	fragment_contacts.xml	联系人页面的布局
	fragment_recent.xml	会话页面的布局
	activity_add_contact.xml	添加联系人页面的布局
	fragment_set.xml	设置页面的布局
	activity_login.xml	登录页面的布局
	activity_welcome.xml	欢迎页面的布局
values	dimens.xml	保存尺寸的 XML 文件
	attrs.xml	保存样式的 XML 文件
	colors.xml	保存颜色的 XML 文件
	strings.xml	保存字符串的 XML 文件
	styles.xml	保存样式的 XML 文件

11.2.3 主要方法流程设计

IM 聊天流程图如图 11-9 所示。

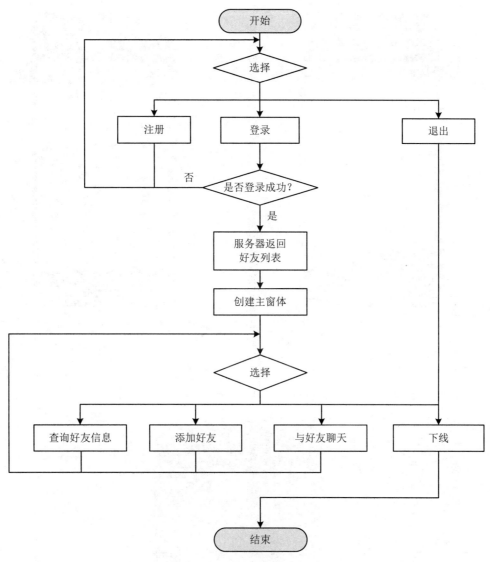

图 11-9 IM 聊天流程图

11.3 代码实现

11.3.1 显示界面布局

1. 系统主界面

系统主界面是进入系统后显示的界面，该界面包括三个 Fragment，如图 11-10 所示。

2. 会话界面

会话界面用于进行好友会话聊天，可以支持发送文字、语音和图片，该界面包括一个 ListView，5 个 Button 和一个 EditText，如图 11-11 所示。

图 11-10　系统主界面　　　　　　图 11-11　会话界面

3. 个人信息界面

个人信息界面用于展示个人信息，可以对头像和昵称进行编辑，该界面包括两个 Button、一个 ImageView 和 7 个 TextView，如图 11-12 所示。

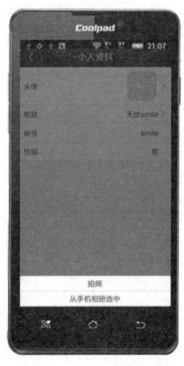

图 11-12　个人信息界面

11.3.2 控件设计实现

1. 带清除功能的输入框

```java
public class ClearEditText extends EditText implements
        OnFocusChangeListener, TextWatcher {
  /**
   * 删除按钮的引用
   */
   private Drawable mClearDrawable;

   public ClearEditText(Context context) {
        this(context, null);
   }

   public ClearEditText(Context context, AttributeSet attrs) {
        //这里构造方法也很重要，不加这个很多属性不能在XML里面定义
        this(context, attrs, android.R.attr.editTextStyle);
   }

   public ClearEditText(Context context, AttributeSet attrs, int defStyle) {
        super(context, attrs, defStyle);
        init();
   }

   private void init() {
        //获取EditText的DrawableRight，假如没有设置我们就使用默认的图片
        mClearDrawable = getCompoundDrawables()[2];
        if (mClearDrawable == null) {
         mClearDrawable = getResources().getDrawable(R.drawable.search_clear);
        }
        mClearDrawable.setBounds(0, 0, mClearDrawable.getIntrinsicWidth(),
            mClearDrawable.getIntrinsicHeight());
        setClearIconVisible(false);
        setOnFocusChangeListener(this);
        addTextChangedListener(this);
   }
/**
     *因为不能直接给EditText设置点击事件，所以用记住按下的位置来模拟点击事件,
     *用输入框的onTouchEvent()方法来模拟。
     *当按下的位置在图标的左边线（EditText的宽度－图标到控件右边的间距－图标的宽度）和图标的
     *右边线（EditText的宽度－图标到控件右边的间距）之间时，就算点击了图标，竖直方向没有考虑。
 */
    @Override
    public boolean onTouchEvent(MotionEvent event) {
        if (getCompoundDrawables()[2] != null) {
```

```java
            if (event.getAction() == MotionEvent.ACTION_UP) {
                boolean touchable = event.getX() > (getWidth()
                        - getPaddingRight() - mClearDrawable.getIntrinsicWidth())
                        && (event.getX() < ((getWidth() - getPaddingRight())));
                if (touchable) {
                    this.setText("");
                }
            }
        }

        return super.onTouchEvent(event);
}

/**
 * 当ClearEditText焦点发生变化的时候，判断里面字符串长度，设置清除图标的显示与隐藏
 */
@Override
public void onFocusChange(View v, boolean hasFocus) {
    if (hasFocus) {
        setClearIconVisible(getText().length() > 0);
    } else {
        setClearIconVisible(false);
    }
}

/**
 * 设置清除图标的显示与隐藏，调用setCompoundDrawables为EditText绘制上去
 * @param visible
 */
protected void setClearIconVisible(boolean visible) {
    Drawable right = visible ? mClearDrawable : null;
    setCompoundDrawables(getCompoundDrawables()[0],
            getCompoundDrawables()[1], right, getCompoundDrawables()[3]);
}

/**
 * 当输入框里内容发生变化的时候回调的方法
 */
@Override
public void onTextChanged(CharSequence s, int start, int count, int after) {
    setClearIconVisible(s.length() > 0);
}

@Override
public void beforeTextChanged(CharSequence s, int start, int count, int after) {

}
```

```java
        @Override
        public void afterTextChanged(Editable s) {

        }


    /**
     * 设置晃动动画
     */
    public void setShakeAnimation(){
        this.setAnimation(shakeAnimation(5));
    }

    /**
     * 晃动动画
     * @param counts表示1秒钟晃动多少下
     * @return
     */
    public static Animation shakeAnimation(int counts){
        Animation translateAnimation = new TranslateAnimation(0, 10, 0, 0);
        translateAnimation.setInterpolator(new CycleInterpolator(counts));
        translateAnimation.setDuration(1000);
        return translateAnimation;
    }
}
```

2. 通讯录右侧快速滚动栏

```java
public class MyLetterView extends View {
    //触摸事件
    private OnTouchingLetterChangedListener onTouchingLetterChangedListener;
    //26个字母
    public static String[] b = { "A", "B", "C", "D", "E", "F", "G", "H", "I",
            "J", "K", "L", "M", "N", "O", "P", "Q", "R", "S", "T", "U", "V",
            "W", "X", "Y", "Z", "#" };
    private int choose = -1;//选中
    private Paint paint = new Paint();

    private TextView mTextDialog;

    public void setTextView(TextView mTextDialog) {
        this.mTextDialog = mTextDialog;
    }

    public MyLetterView(Context context, AttributeSet attrs, int defStyle) {
        super(context, attrs, defStyle);
    }
```

```java
public MyLetterView(Context context, AttributeSet attrs) {
    super(context, attrs);
}

public MyLetterView(Context context) {
    super(context);
}

/**
 * 重写这个方法
 */
protected void onDraw(Canvas canvas) {
    super.onDraw(canvas);
    //获取焦点改变背景颜色
    int height = getHeight();              //获取对应高度
    int width = getWidth();                //获取对应宽度
    int singleHeight = height / b.length;  //获取每一个字母的高度

    for (int i = 0; i < b.length; i++) {
        paint.setColor(getResources().getColor(R.color.color_bottom_text_normal));
        paint.setTypeface(Typeface.DEFAULT_BOLD);
        paint.setAntiAlias(true);
        paint.setTextSize(PixelUtil.sp2px(12));
        //选中的状态
        if (i == choose) {
            paint.setColor(Color.parseColor("#3399ff"));
            paint.setFakeBoldText(true);
        }
        //x坐标等于中间-字符串宽度的一半
        float xPos = width / 2 - paint.measureText(b[i]) / 2;
        float yPos = singleHeight * i + singleHeight;
        canvas.drawText(b[i], xPos, yPos, paint);
        paint.reset();                     //重置画笔
    }

}

@SuppressWarnings("deprecation")
@Override
public boolean dispatchTouchEvent(MotionEvent event) {
    final int action = event.getAction();
    final float y = event.getY();              //点击y坐标
    final int oldChoose = choose;
    final OnTouchingLetterChangedListener listener = onTouchingLetterChangedListener;
    //点击y坐标所占总高度的比例*b数组的长度就等于点击b中的个数
    final int c = (int) (y / getHeight() * b.length);
```

```java
        switch (action) {
        case MotionEvent.ACTION_UP:
            setBackgroundDrawable(new ColorDrawable(0x00000000));
            choose = -1;//
            invalidate();
            if (mTextDialog != null) {
                mTextDialog.setVisibility(View.INVISIBLE);
            }
            break;

        default:
            //设置右侧字母列表[A,B,C,D,E....]的背景颜色
            setBackgroundResource(R.drawable.v2_sortlistview_sidebar_background);
            if (oldChoose != c) {
                if (c >= 0 && c < b.length) {
                    if (listener != null) {
                        listener.onTouchingLetterChanged(b[c]);
                    }
                    if (mTextDialog != null) {
                        mTextDialog.setText(b[c]);
                        mTextDialog.setVisibility(View.VISIBLE);
                    }

                    choose = c;
                    invalidate();
                }
            }

            break;
        }
        return true;
    }

    /**
     * 向外公开的方法
     *
     * @param onTouchingLetterChangedListener
     */
    public void setOnTouchingLetterChangedListener(
            OnTouchingLetterChangedListener onTouchingLetterChangedListener) {
        this.onTouchingLetterChangedListener = onTouchingLetterChangedListener;
    }

    /**
     *接口
     *
     * @author coder
```

```java
     *
     */
    public interface OnTouchingLetterChangedListener {
        public void onTouchingLetterChanged(String s);
    }

}
```

3. 提示对话框

```java
public class DialogTips extends DialogBase {
    boolean hasNegative;
    boolean hasTitle;
    /**
     * 构造函数
     * @param context
     */
    public DialogTips(Context context, String title,String message,String buttonText,boolean
            hasNegative,boolean hasTitle) {
        super(context);
        super.setMessage(message);
        super.setNamePositiveButton(buttonText);
        this.hasNegative = hasNegative;
        this.hasTitle = hasTitle;
        super.setTitle(title);
    }

    /**下线通知的对话框样式
     * @param context
     * @param title
     * @param message
     * @param buttonText
     */
    public DialogTips(Context context,String message,String buttonText) {
        super(context);
        super.setMessage(message);
        super.setNamePositiveButton(buttonText);
        this.hasNegative = false;
        this.hasTitle = true;
        super.setTitle("提示");
        super.setCancel(false);
    }

    public DialogTips(Context context, String message,String buttonText,String negetiveText,String
            title,boolean isCancel) {
        super(context);
        super.setMessage(message);
        super.setNamePositiveButton(buttonText);
```

```java
        this.hasNegative=false;
        super.setNameNegativeButton(negetiveText);
        this.hasTitle = true;
        super.setTitle(title);
        super.setCancel(isCancel);
    }

    /**
     *创建对话框
     */
    @Override
    protected void onBuilding() {
        super.setWidth(dip2px(mainContext, 300));
        if(hasNegative){
            super.setNameNegativeButton("取消");
        }
        if(!hasTitle){
            super.setHasTitle(false);
        }
    }

    public int dip2px(Context context,float dipValue){
        float scale=context.getResources().getDisplayMetrics().density;
        return (int) (scale*dipValue+0.5f);
    }

    @Override
    protected void onDismiss() { }

    @Override
    protected void OnClickNegativeButton() {
        if(onCancelListener != null){
            onCancelListener.onClick(this, 0);
        }
    }

    /**
     * 确认按钮，触发onSuccessListener的OnClick
     */
    @Override
    protected boolean OnClickPositiveButton() {
        if(onSuccessListener != null){
            onSuccessListener.onClick(this, 1);
        }
        return true;
    }
}
```

4. 支持表情显示的文本框

```java
public class EmoticonsTextView extends TextView {

    public EmoticonsTextView(Context context) {
        super(context);
    }

    public EmoticonsTextView(Context context, AttributeSet attrs, int defStyle) {
        super(context, attrs, defStyle);
    }

    public EmoticonsTextView(Context context, AttributeSet attrs) {
        super(context, attrs);
    }

    @Override
    public void setText(CharSequence text, BufferType type) {
        if (!TextUtils.isEmpty(text)) {
            super.setText(replace(text.toString()), type);
        } else {
            super.setText(text, type);
        }
    }

    private Pattern buildPattern() {
        return Pattern.compile("\\\\ue[a-z0-9]{3}", Pattern.CASE_INSENSITIVE);
    }

    private CharSequence replace(String text) {
        try {
            SpannableString spannableString = new SpannableString(text);
            int start = 0;
            Pattern pattern = buildPattern();
            Matcher matcher = pattern.matcher(text);
            while (matcher.find()) {
                String faceText = matcher.group();
                String key = faceText.substring(1);
                BitmapFactory.Options options = new BitmapFactory.Options();
                Bitmap bitmap = BitmapFactory.decodeResource(getContext().getResources(),
                        getContext().getResources().getIdentifier(key, "drawable",
                        getContext().getPackageName()), options);
                ImageSpan imageSpan = new ImageSpan(getContext(), bitmap);
                int startIndex = text.indexOf(faceText, start);
                int endIndex = startIndex + faceText.length();
                if (startIndex >= 0)
                    spannableString.setSpan(imageSpan, startIndex, endIndex,
```

```
                    Spannable.SPAN_EXCLUSIVE_EXCLUSIVE);
                start = (endIndex - 1);
            }
            return spannableString;
        } catch (Exception e) {
            return text;
        }
    }
  }
}
```

11.3.3 Socket 线程

```
public class SocketThread{
    private Socket socket;
    private String ip,port;
    private InetSocketAddress isa;
    private DataOutputStream DOS=null;
    private DataInputStream DIS=null;

    SocketMap smMap;

    public void SocketStart(String myip,String myport,String type){
        this.ip=myip;
        this.port=myport;

        //new Thread(){
            //public void run(){
                socket = new Socket();
                isa = new InetSocketAddress(ip,Integer.parseInt(port));
                try {
                    socket.connect(isa,5000);
                    System.out.println("连接成功"+socket);
                    smMap = new SocketMap();
                    smMap.setSocket(type, socket);
                }catch (SocketTimeoutException e) {
                    e.printStackTrace();
                    System.out.println("连接超时"+e.toString());
                } catch (IOException e) {
                    e.printStackTrace();
                    System.out.println("连接失败"+e.toString());
                }
            //}
        //}.start();
    }

    public DataOutputStream getDOS() throws IOException{
        DOS = new DataOutputStream (this.socket.getOutputStream());
```

```
        return DOS;
    }

    public DataInputStream getDIS() throws IOException{
        //System.out.println("DIS:"+socket);
        DIS = new DataInputStream (this.socket.getInputStream());
        return DIS;
    }

    public Socket getSocket(){
        return this.socket;
    }

    public void setIP(String setip){
        this.ip=setip;
    }
    public void setPort(String setport){
        this.port=setport;
    }

    public boolean isConnected(){
        return socket.isConnected();
    }

    public void CloseSocket(String type){
        smMap.removeMap(type);
    }

    public void AllClose(){
        smMap.clearMap();
    }
}
```

11.3.4 待发消息队列

```
public class MsgRequestQueueManager {

    LinkedBlockingQueue<MsgRequest> mQueueList;

    private static final class getInstance{
        public static final MsgRequestQueueManager mMsgRequestQueueManager = new
            MsgRequestQueueManager();
    }

    private MsgRequestQueueManager(){
        mQueueList = new LinkedBlockingQueue<MsgRequest>();
```

```
    }

    public static MsgRequestQueueManager getInstance(){
        return getInstance.mMsgRequestQueueManager;
    }

    public MsgRequest poll(){
        MsgRequest entity = null;
        if(mQueueList != null){
            entity =   mQueueList.poll();
        }
        return entity;
    }

    public void push(MsgRequest entity){
        try {
            if(mQueueList != null){
                mQueueList = new LinkedBlockingQueue<MsgRequest>();
            }
            mQueueList.put(entity);
        } catch (InterruptedException e) {
            e.printStackTrace();
        }
    }

}
```

11.3.5 消息接收

```
package com.jpy.myapplication;

import android.text.TextUtils;

import java.io.BufferedInputStream;
import java.io.BufferedOutputStream;
import java.io.DataInputStream;
import java.io.File;
import java.io.FileNotFoundException;
import java.io.FileOutputStream;
import java.io.IOException;
import java.io.InputStream;
import java.net.Socket;
import java.net.SocketException;

public class MsgReceiveHandler extends Thread {

    private static final int TYPE_FILE = 0;
```

```java
private static final int TYPE_TEXT = 1;
InputStream mInputStream = null;
DataInputStream mDataInputStream = null;

private Socket mSocket;

public MsgReceiveHandler(Socket socket) {
    this.mSocket = socket;
}

public void run() {

    try {
        mInputStream = mSocket.getInputStream();

        mDataInputStream = new DataInputStream(new BufferedInputStream(
                mInputStream));

        while (true) {

            try {
                Thread.sleep(1000);
            } catch (InterruptedException e) {
                e.printStackTrace();
            }

            if (mSocket == null || mDataInputStream == null) {
                return;
            }

            int type = mDataInputStream.readInt();

            if (type == TYPE_FILE) {
                // 接收文件
                doFileReceive(mDataInputStream);
            } else if(type == TYPE_TEXT){
                doTextReceive(mDataInputStream);
            }

        }

    } catch (SocketException e1) {
        e1.printStackTrace();
    } catch (IOException e1) {
        e1.printStackTrace();
    }
```

}

/**
 * 接收文本信息
 * @param dis
 * @throws IOException
 */
private void doTextReceive(DataInputStream dis) throws IOException {
    String msg = null;
    int length = (int)dis.readInt();

    byte[] buffer = new byte[length];
    dis.read(buffer, 0, buffer.length);
    msg = new String(buffer);

    if (!TextUtils.isEmpty(msg.trim())) {
        //处理接收的文本信息
    }
}

/**
 * 接收文件处理
 *
 * @param dis
 */
private void doFileReceive(DataInputStream dis){

    if (dis == null) {
        return;
    }

    BufferedOutputStream bufferedOutputStream = null;

    try {

        // 文件名
        String fileName = dis.readUTF();

        // 存储路径
        if (TextUtils.isEmpty(fileName)) {
            return;
        }

        long size = dis.readLong();
        if (size <= 0) {
            return;

```java
                }

                int time = dis.readInt();

                String filePath = getFilePath(fileName);

                // 创建目录
                if(CreateDir(filePath)){
                    bufferedOutputStream = new BufferedOutputStream(new FileOutputStream(new File(filePath)));
                }

                int readlen = 0;
                byte[] buffer = new byte[1024];
                int writeLens = 0;
                while ((readlen = dis.read(buffer, 0, buffer.length)) != -1) {
                    writeLens += readlen;

                    if(bufferedOutputStream != null){
                        bufferedOutputStream.write(buffer, 0, readlen);
                    }
                    // 如果文件读取完，就退出循环，避免阻塞
                    if (writeLens >= size) {
                        break;
                    }
                }
                if(bufferedOutputStream != null){
                    bufferedOutputStream.flush();

                }
        } catch (FileNotFoundException e) {
            e.printStackTrace();
        } catch (IOException e) {
            // e.printStackTrace();

        } finally {
            try {
                if (bufferedOutputStream != null) {
                    bufferedOutputStream.close();
                }
            } catch (IOException e) {
                e.printStackTrace();
            }
        }
    }

}
```

11.4 关键知识点解析

11.4.1 Socket 定义

Socket 即套接字，它不是一种协议，而是一个编程调用接口，属于传输层（主要解决数据如何在网络中传输），通过 Socket 用户可以在 Android 平台上通过 TCP/IP 协议进行开发。

Socket 套接字成对出现，代码格式如下：

Socket ={(IP地址1:PORT端口号), (IP地址2:PORT端口号)}

一个 Socket 实例唯一代表一个主机上的一个应用程序的通信链路。

Socket 的使用类型主要有以下两种：

- 流套接字（StreamSocket）：基于 TCP 协议，采用流的方式提供可靠的字节流服务。
- 数据报套接字(DatagramSocket)：基于 UDP 协议，采用数据报文提供数据打包发送的服务。

11.4.2 Socket 与 HTTP 对比

Socket 属于传输层，因为 TCP/IP 协议属于传输层，解决的是数据如何在网络中传输的问题。

而 HTTP 协议属于应用层，解决的是如何包装数据的问题。二者不属于同一层面，本来是没有可比性的。但随着发展，默认的 HTTP 里封装了下面几层的使用，所以才会出现 Socket 与 HTTP 协议的对比（主要是工作方式的不同）。

（1）HTTP 采用请求－响应方式，即建立网络连接后，只有当客户端向服务器发送请求，服务端才能向客户端返回数据，一问一答形式只有使用才建立连接。可理解为客户端有需要才进行通信。

（2）Socket 采用服务器主动发送数据的方式，即建立网络连接后，服务器可主动发送消息给客户端，而不需要由客户端向服务器发送请求。需要一直保存连接。可理解为服务端有需要才进行通信。

11.4.3 使用 UDP 协议通信

```
protected void connectServerWithUDPSocket() {

    DatagramSocket socket;
    try {
        //创建DatagramSocket对象并指定一个端口号
        socket = new DatagramSocket(1111);
        // IP地址转换为网络地址
        InetAddress serverAddress = InetAddress.getByName("192.168.0.7");

        String str = "要发送的报文";        //设置要发送的报文
        byte data[] = str.getBytes();       //把字符串str转换为字节数组
        //创建一个DatagramPacket对象，用于发送数据
```

```
            DatagramPacket packet = new DatagramPacket(data, data.length ,serverAddress ,1005);
//把数据发送到服务端
            socket.send(packet);
        } catch (SocketException e) {
            e.printStackTrace();
        } catch (UnknownHostException e) {
            e.printStackTrace();
        } catch (IOException e) {
            e.printStackTrace();
        }
    }
```

11.5　问题与讨论

1．Socket 通信是否需要放在工作线程中？
2．如何解决手机锁屏 Socket 长连接挂起问题？
3．如何进行客户端服务端都判断是否存活？
4．如何解决 Socket 自动拆包问题？

项目 12　易行打车

本项目设计一款仿照"滴滴打车"的打车软件,其中包括模块技术、多线程、百度地图、意见反馈、登录注册、在线更新等模块功能。

项目需求描述如下:
(1)用户注册和登录模块。
(2)打车模块,功能包括选择始发地和目的地、是否拼车、进行打车呼叫。
(3)百度地图模块,功能包括显示地图、定位和地图标注。
(4)意见反馈模块,填写反馈信息并进行提交。
(5)在线更新模块,功能包括版本检查和新版本下载更新。

(1)学习百度地图。
(2)学习在线更新。

12.1　总体设计

12.1.1　总体分析

通过百度地图实现打车功能,包含打车、查看打车记录和客户端在线更新。

12.1.2　功能模块框图

根据总体分析结果可以总结出功能模块框图,如图 12-1 所示。

图 12-1　功能模块框图

总体模块的作用主要是生成应用程序的主类，控制应用程序的生命周期；基础架构模块主要提供程序架构、所有 Activity 公用的父类、所有 Activity 公用的方法，包括自定义风格对话框、自定义提示框等功能；数据管理与控制模块主要提供数据获取、数据解析、数据读写、数据组织和数据缓存功能；用户界面模块包括登录、打车、历史记录、设置、操作提示等功能；网络通信模块主要负责客户端与服务端的数据交互，包括客户端从服务端获取打车历史记录等数据信息，客户端向服务端提交登录、打车请求数据等信息。

12.1.3 系统流程图

根据总体分析结果及功能模块框图梳理出系统启动的主要流程，如图 12-2 所示。

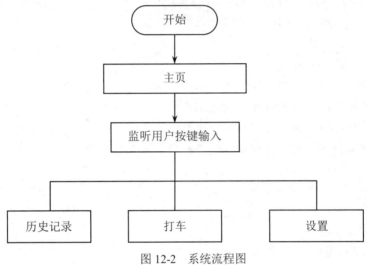

图 12-2 系统流程图

12.1.4 界面设计

本应用是打车应用，主要实现打车功能。
根据程序功能需求可以规划出软件的主要界面，如下：
（1）打车：显示地图、选择始发地、目的地和乘车人数。
（2）设置菜单：登录、注册、历史记录、设置、切换账号、退出。
系统主界面如图 12-3 所示。

图 12-3 系统主界面

从图 12-3 中可以很直观地看到，主界面包含叫车和设置两个功能。

12.2 详细设计

12.2.1 模块描述

在系统总体分析及界面布局设计完成后，主要工作就转入对各个功能模块的详细设计阶段。

1. 基础架构模块详细设计

基础架构模块主要提供程序架构、所有 Activity 公用的父类、所有 Activity 公用的方法，包括自定义风格对话框、自定义提示框等功能。

基础架构模块功能如图 12-4 所示。

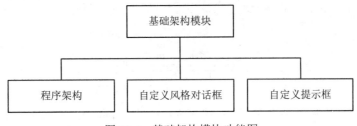

图 12-4　基础架构模块功能图

2. 用户界面模块详细设计

用户界面模块的主要任务是显示打车界面和实现与用户的交互，即当用户点击按键或者屏幕的时候，监听器会去调用相应的处理办法或其他相应的处理模块。

本模块包括叫车、打车记录和设置等功能。

用户界面模块功能如图 12-5 所示。

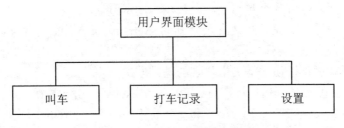

图 12-5　用户界面模块功能图

3. 数据管理与控制模块详细设计

数据管理与控制模块主要提供数据获取、数据解析、数据读写、数据组织和数据缓存功能。

数据管理与控制模块和用户界面模块可以调用基础架构模块的一些通用方法。数据管理与控制模块为用户界面模块提供数据，同时可以接收并保存用户界面模块产生的数据。

数据管理与控制模块功能如图 12-6 所示。

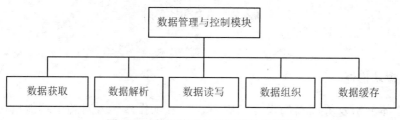

图 12-6　数据管理与控制模块功能图

4. 网络通信模块详细设计

网络通信模块根据用户界面的需求调用访问服务器，接收服务器返回的数据并解析，同时显示到用户界面。

本模块包括发送网络请求、接收网络应答、网络数据解析等功能。

网络通信模块功能如图 12-7 所示。

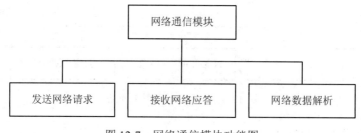

图 12-7　网络通信模块功能图

12.2.2　系统包及其资源规划

1. 文件结构

在系统各个模块的实现方式和流程设计完成后，就可以对系统主要的包和资源进行规划，划分的原则主要是保持各个包相互独立，耦合度尽量低。

根据系统功能设计，本系统封装一个基础的 Activity 类，加载各个子页的通用控件，并提供一些基础的实现方法，例如设置进度条、标题等常用的方法，程序中的 Activity 都可继承此基类，继承后就可直接使用基类中封装的基础方法。

系统使用 11 个 Activity 和 3 个 Fragment：一个 Activity 用于注册，一个 Activity 用于登录，一个 Activity 用于主页，3 个 Activity 用于打车，一个 Activity 用于叫车记录，一个 Activity 用于叫车成功，一个 Activity 用于设置，一个 Activity 用于修改密码，一个 Activity 用于修改姓名。包及其资源结构如图 12-8 所示。

2. 命名空间

本示例设置了多个命名空间，分别用来保存用户界面、后台服务的源代码文件，具体说明见表 12-1。

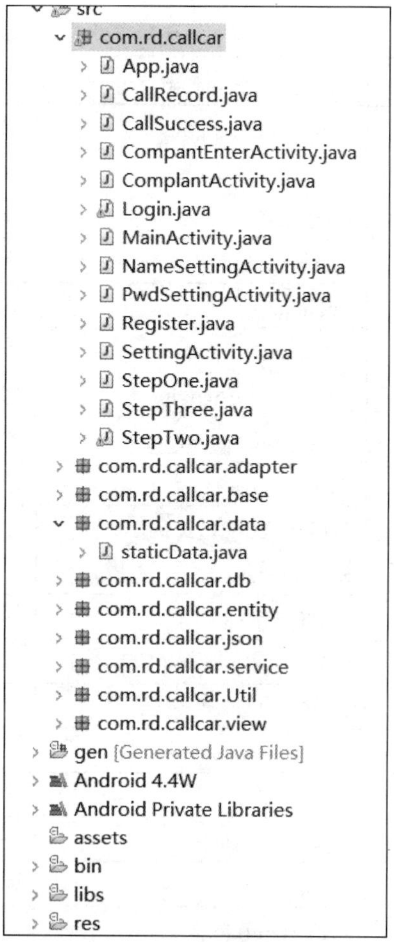

```
v 🗁 src
  v ⊞ com.rd.callcar
    > 🗋 App.java
    > 🗋 CallRecord.java
    > 🗋 CallSuccess.java
    > 🗋 CompantEnterActivity.java
    > 🗋 ComplantActivity.java
    > 🗋 Login.java
    > 🗋 MainActivity.java
    > 🗋 NameSettingActivity.java
    > 🗋 PwdSettingActivity.java
    > 🗋 Register.java
    > 🗋 SettingActivity.java
    > 🗋 StepOne.java
    > 🗋 StepThree.java
    > 🗋 StepTwo.java
  > ⊞ com.rd.callcar.adapter
  > ⊞ com.rd.callcar.base
  v ⊞ com.rd.callcar.data
    > 🗋 staticData.java
  > ⊞ com.rd.callcar.db
  > ⊞ com.rd.callcar.entity
  > ⊞ com.rd.callcar.json
  > ⊞ com.rd.callcar.service
  > ⊞ com.rd.callcar.Util
  > ⊞ com.rd.callcar.view
> 🗁 gen [Generated Java Files]
> 🗁 Android 4.4W
> 🗁 Android Private Libraries
  🗁 assets
> 🗁 bin
> 🗁 libs
> 🗁 res
```

图 12-8 包及其资源结构

表 12-1 命名空间

命名空间	说明
com.rd.callcar	存放与页面视图相关的源代码文件
com.rd.callcar.adapter	存放与页面适配器相关的源代码文件
com.rd.callcar.base	存放与页面视图相关的源代码文件
com.rd.callcar.data	存放与配置相关的源代码文件
com.rd.callcar.entity	存放与数据对象相关的源代码文件
com.rd.callcar.service	存放与自动更新相关的源代码文件
com.rd.callcar.Util	存放与工具相关的源代码文件
com.rd.callcar.view	存放与自定义 View 相关的源代码文件

3. 源代码文件

源代码文件及说明见表 12-2。

表 12-2 源代码文件

包名称	文件名	说明
com.rd.callcar	CallRecord.java	叫车记录页面
	CallSuccess.java	叫车成功页面
	CompantEnterActivity.java	反馈页面
	ComplantActivity.java	投诉页面
	Login.java	登录页面
	MainActivity.java	主页面
	NameSettingActivity	修改姓名页面
	PwdSettingActivity.java	修改密码页面
	Register.java	注册页面
	SettingActivity.java	设置页面
	StepOne.java	叫车页面 1
	StepTwo.java	叫车页面 2
com.rd.callcar.adapter	ChooseAdapter.java	选择适配器类
	ComplantAdapter.java	反馈适配器类
	HistoryAdapter.java	叫车历史列表适配器类
com.rd.callcar.entity	User.java	个人信息类
	FaceText.java	表情类
com.rd.callcar.data	staticData.java	常量配置类
com.rd.callcar.Util	getSystemInfo.java	获取系统信息工具类
	ExitApplication.java	Activity 工具类
	UpdateCustomer.java	更新工具类
com.rd.callcar.entity	CallHistory.java	叫车历史对象类
	PointInfo.java	地图绘点对象类
	UpdataInfo.java	更新对象类
	FeedMsg.java	反馈信息对象类
com.rd.callcar.service	UpdateService.java	自动更新服务
com.rd.callcar.view	MarqueeText.java	跑马灯文本框

4. 资源文件

Android 的资源文件保存在/res 的子目录中。

- /res/drawable 目录：保存的是图像文件。
- /res/layout 目录：保存的是布局文件。
- /res/values 目录：保存的是用来定义字符串和颜色的文件。

资源文件及说明见表 12-3。

表 12-3 资源文件

资源目录	文件	说明
drawable	splash.png	欢迎页背景
	chat_fail_resend_normal.png	消息发送失败图标
	chat_keyboard_normal.png	键盘按钮
	ic_launcher.png	程序图标文件
	bg.jpg	主界面背景图
	chat_voice_normal.png	语音按钮
	default_head.png	默认头像
	icon_near.png	位置图标
	list_newmessage2.9.png	消息背景
	msg_tips.png	新信息提示
	ic_bluetooth.png	蓝牙图标
	new_friends_icon.png	添加好友按钮
	refresh_black.png	刷新
layout	activity_main.xml	主界面的布局
	fragment_contacts.xml	联系人页面的布局
	fragment_recent.xml	会话页面的布局
	activity_add_contact.xml	添加联系人页面的布局
	fragment_set.xml	设置页面的布局
	activity_login.xml	登录页面的布局
	activity_welcome.xml	欢迎页面的布局
values	dimens.xml	保存尺寸的 XML 文件
	attrs.xml	保存样式的 XML 文件
	colors.xml	保存颜色的 XML 文件
	strings.xml	保存字符串的 XML 文件
	styles.xml	保存样式的 XML 文件

12.2.3 主要方法流程设计

打车流程图如图 12-9 所示。

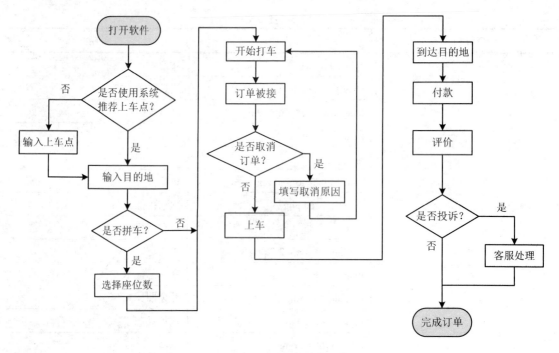

图 12-9　打车流程图

12.3　代码实现

12.3.1　显示界面布局

1. 系统主界面

系统主界面是进入系统后显示的界面，该界面包括 3 个 Button、两个 TextView、一个 EditText 和一个 Menu，如图 12-10 所示。

图 12-10　系统主界面

2. 叫车设置界面

叫车设置界面可以对终点站和乘车人数进行设置，该界面包括两个 TextView、一个 Spinner、一个 Button 和一个 EditText，如图 12-11 所示。

图 12-11　叫车设置界面

3. 设置界面

设置界面功能包括姓名昵称编辑、密码编辑、关于软件和检查更新等，该界面包括 4 个 Button 和 8 个 TextView，如图 12-12 所示。

图 12-12　设置界面

12.3.2 控件设计实现

1. 自定义 marker 覆盖类

```java
public class CustomOverlayItem extends ItemizedOverlay<OverlayItem> {
    //private List<OverlayItem> GeoList = new ArrayList<OverlayItem>();
    private Context mContext;
    private OverlayItem overlay;
    boolean showtext;
    //private String title;
    private Drawable marker;

    /**
     * marker构造类
     *
     * @param marker
     * @param context
     * @param p
     * @param title
     * @param sinppet
     * @param showtext
     */
    public CustomOverlayItem(Drawable marker, Context context, GeoPoint p,
            String title, String sinppet, boolean showtext) {
        super(boundCenterBottom(marker));
        this.mContext = context;
        //用给定的经纬度构造GeoPoint，单位是微度（度*1E6）
        //point = p;
        this.showtext = showtext;
        //this.title = title;
        this.marker = marker;
        overlay = new OverlayItem(p, title, sinppet);
        populate();    //createItem(int)方法构造item。一旦有了数据，在调用其他方法前首先调用这个方法
    }

    @Override
    protected OverlayItem createItem(int i) {
        return overlay;
    }

    @Override
    public int size() {
        return 1;
    }

    @Override
    public void draw(Canvas canvas, MapView mapView, boolean arg2) {
```

```java
        super.draw(canvas, mapView, arg2);
        //Projection接口用于屏幕像素坐标和经纬度坐标之间的变换
        Projection projection = mapView.getProjection();
        String title = overlay.getTitle();
        //把经纬度变换到相对于MapView左上角的屏幕像素坐标
        Point point = projection.toPixels(overlay.getPoint(), null);
        //可在此处添加您的绘制代码
        Paint paintText = new Paint();
        Paint paint = new Paint();
        paint.setAlpha(255);
        paint.setColor(Color.DKGRAY);
        paint.setStrokeWidth(5);
        paintText.setColor(Color.BLUE);
        paintText.setTextSize(15);
        //canvas.drawCircle(point.x, point.y, 100, paint);
        canvas.drawText(title, point.x - 30, point.y - 50, paintText);     //绘制文本
        //调整一个drawable边界，使得(0,0)是这个drawable底部最后一行中心的一个像素
        boundCenterBottom(marker);
    }

    @Override
    //当点击事件时处理
    protected boolean onTap(int i) {
        if (showtext)
            Toast.makeText(this.mContext, overlay.getTitle(), Toast.LENGTH_SHORT).show();
        return true;
    }
}
```

2. 自定义路线图类

```java
public class CustomRouteOverLay extends RouteOverlay {

    public Activity ac;
    private MapView mapView;

    static ArrayList<View> overlayviews = new ArrayList<View>();

    public CustomRouteOverLay(Activity arg0, MapView arg1) {
        super(arg0, arg1);
        ac = arg0;
        mapView = arg1;
    }

    @Override
    protected boolean onTap(int arg0) {
        //return super.onTap(arg0);
        return true;
```

```java
    }

    @Override
    public void setData(MKRoute arg0) {
        super.setData(arg0);
        addHint(arg0);
    }

    public void addHints(MKRoute routes) {
        for (int i = 0; i < routes.getNumSteps(); i++) {
            Drawable marker = ac.getResources().getDrawable(R.drawable.pop);   //得到需要标在地图上的资源
            marker.setBounds(0, 0, marker.getIntrinsicWidth(),
                    marker.getIntrinsicHeight());          //为marker定义位置和边界
            OverItemT overitem = new OverItemT(marker, ac, routes.getStep(i)
                    .getContent(), routes.getStep(i).getPoint());
            //OverlayItem over=new OverlayItem(routes.GET, null, null);
            mapView.getOverlays().add(overitem);          //添加ItemizedOverlay实例到mMapView
        }
        mapView.invalidate();
    }

    /**
     * 增加指示路线
     *
     * @param routes
     */
    public void addHint(MKRoute routes) {
        mapView.getOverlays().clear();              //先清空
        //mapView.removeAllViewsInLayout();
        View mPopView = ac.getLayoutInflater().inflate(R.layout.popview, null);
        for (int i = 0; i < overlayviews.size(); i++) {
            System.out.println("remove &" + i);
            mapView.removeViewInLayout(overlayviews.get(i));
            overlayviews.remove(i);
        }

        mapView.invalidate();
        //添加ItemizedOverlay
        for (int i = 0; i < routes.getNumSteps(); i++) {

            Drawable marker = ac.getResources().getDrawable(R.drawable.pop); //得到需要标在地图上的资源
            marker.setBounds(0, 0, marker.getIntrinsicWidth(),
                    marker.getIntrinsicHeight());          //为marker定义位置和边界
            GeoPoint pt = routes.getStep(i).getPoint();    //=routes.get(i).getPoint();
            if (i != 0 && i != routes.getNumSteps() - 1) {
                mPopView = ac.getLayoutInflater().inflate(R.layout.popview,null);
```

```java
            mapView.addView(mPopView, new MapView.LayoutParams(
                    LayoutParams.WRAP_CONTENT, LayoutParams.WRAP_CONTENT,
                    null, MapView.LayoutParams.TOP_LEFT));
            mPopView.setVisibility(View.GONE);
            mapView.updateViewLayout(mPopView, new MapView.LayoutParams(
                    LayoutParams.WRAP_CONTENT, LayoutParams.WRAP_CONTENT,
                    pt, MapView.LayoutParams.BOTTOM_CENTER));
            mPopView.setVisibility(View.VISIBLE);
            Button button = (Button) mPopView.findViewById(R.id.overlay_pop);
            button.setText(routes.getStep(i).getContent());
            overlayviews.add(mPopView);
            overlayviews.add(button);
        } else {
            //修改起始点和终点样式（自定义）
            mPopView = ac.getLayoutInflater().inflate(R.layout.popview,null);
            mapView.addView(mPopView, new MapView.LayoutParams(
                    LayoutParams.WRAP_CONTENT, LayoutParams.WRAP_CONTENT,
                    null, MapView.LayoutParams.TOP_LEFT));
            mPopView.setVisibility(View.GONE);
            mapView.updateViewLayout(mPopView, new MapView.LayoutParams(
                    LayoutParams.WRAP_CONTENT, LayoutParams.WRAP_CONTENT,
                    pt, MapView.LayoutParams.BOTTOM_CENTER));
            mPopView.setVisibility(View.VISIBLE);
            Button button = (Button) mPopView.findViewById(R.id.overlay_pop);
            button.offsetTopAndBottom(100);
            button.setTextColor(Color.BLUE);
            button.setBackgroundColor(Color.TRANSPARENT);
            button.setText(routes.getStep(i).getContent());
            overlayviews.add(mPopView);
            overlayviews.add(button);
        }
    }
}

class OverItemT extends ItemizedOverlay<OverlayItem> {

    private Drawable marker;
    private OverlayItem o;

    public OverItemT(Drawable marker, Context context, String title,GeoPoint p) {
        super(boundCenterBottom(marker));
        this.marker = marker;
        //构造OverlayItem的三个参数依次为item的位置、标题文本和文字片段
        o = new OverlayItem(p, title, title);
        //createItem(int)方法构造item。一旦有了数据，在调用其他方法前首先调用这个方法
        populate();
```

```java
    }

    public void updateOverlay() {
        populate();
    }

    @Override
    public void draw(Canvas canvas, MapView mapView, boolean shadow) {

        //Projection接口用于屏幕像素坐标和经纬度坐标之间的变换
        Projection projection = mapView.getProjection();
        for (int index = size() - 1; index >= 0; index--) { //遍历mGeoList
            OverlayItem overLayItem = getItem(index); //得到给定索引的item
            String title = overLayItem.getTitle();
            //把经纬度变换到相对于MapView左上角的屏幕像素坐标
            Point point = projection.toPixels(overLayItem.getPoint(), null);
            //可在此处添加您的绘制代码
            Paint paintText = new Paint();
            paintText.setColor(Color.BLUE);
            paintText.setTextSize(15);
            canvas.drawText(title, point.x-30, point.y, paintText);      //绘制文本
        }
        super.draw(canvas, mapView, shadow);
        //调整一个drawable边界，使得(0,0)是这个drawable底部最后一行中心的一个像素
        boundCenterBottom(marker);
    }

    @Override
    protected OverlayItem createItem(int i) {
        return o;
    }

    @Override
    public int size() {
        return 1;
    }

    @Override
    //当点击事件时处理
    protected boolean onTap(int i) {
        //更新气泡位置并使之显示
        return true;
    }

    @Override
    public boolean onTap(GeoPoint arg0, MapView arg1) {
```

```java
            //消去弹出的气泡
            //ItemizedOverlayDemo.mPopView.setVisibility(View.GONE);
            return super.onTap(arg0, arg1);
        }
    }
}
```

3. 自定对话框类

```java
public class DialogManager {
    public static void showExitHit(final Activity act, int title, int msg) {
        final AlertDialog alertDialog = new AlertDialog.Builder(act)
                .setTitle(title)
                .setMessage(msg)
                .setPositiveButton(R.string.sure,new DialogInterface.OnClickListener() {
                        public void onClick(DialogInterface dialog,int which) {
                            act.finish();
                        }
                    })
                .setNegativeButton(R.string.cancel, new DialogInterface.OnClickListener() {
                        public void onClick(DialogInterface dialog, int which) {
                            return;
                        }
                    }).create();
        alertDialog.show();
    }

    public void showExitHit(final Activity act) {
        final AlertDialog alertDialog = new AlertDialog.Builder(act)
                .setTitle(R.string.exitHint)
                .setMessage(R.string.exitChoose)
                .setPositiveButton(R.string.sure,new DialogInterface.OnClickListener() {
                        public void onClick(DialogInterface dialog,int which) {
                            ExitApplication.getInstance().exit();
                        }
                    })
                .setNegativeButton(R.string.cancel,
                        new DialogInterface.OnClickListener() {
                            public void onClick(DialogInterface dialog,int which) {
                                return;
                            }
                        }).create();
        alertDialog.show();
    }

    public void toggleGPS(Activity act) {
        Intent gpsIntent = new Intent();
```

```
            gpsIntent.setClassName("com.android.settings",
                "com.android.settings.widget.SettingsAppWidgetProvider");
            gpsIntent.addCategory("android.intent.category.ALTERNATIVE");
            gpsIntent.setData(Uri.parse("custom:3"));
            try {
                PendingIntent.getBroadcast(act, 0, gpsIntent, 0).send();
            } catch (CanceledException e) {
                e.printStackTrace();
            }
        }
    }
}
```

12.3.3 申请百度地图 API Key

要想使用百度地图，首先要申请 Key，目前百度地图 Android 平台 API 申请链接为 http://lbsyun.baidu.com/apiconsole/key，如图 12-13 所示。

图 12-13 百度地图 Android 平台 API 申请界面

创建应用，填写包名和安全码，如图 12-14 所示。

图 12-14 信息填写界面

12.3.4 初始化定位

```
/**
 * 初始化定位
 *
 * @param context
 */
public void initLocation(Context context) {

    mLocationClient = new LocationClient(context);
    mMyLocationListener = new MyLocationListener();
    mLocationClient.registerLocationListener(mMyLocationListener);
    initLocation();
    mLocationClient.start();    //定位SDK
    //start之后会默认发起一次定位请求，开发者无须判断isstart并主动调用request
    mLocationClient.requestLocation();
}

private void initLocation() {
    LocationClientOption option = new LocationClientOption();

    option.setLocationMode(tempMode);    //可选，默认高精度，设置定位模式，高精度，低功耗，仅设备
    //option.setCoorType(tempcoor);      //可选，默认gcj02，设置返回的定位结果坐标系

    //option.setScanSpan(span);          //可选，默认0，即仅定位一次，设置发起定位请求的间隔需要大于
                                         //等于1000ms才是有效的
    option.setIsNeedAddress(true);       //可选，设置是否需要地址信息，默认不需要
    //option.setOpenGps(true);           //可选，默认false，设置是否使用GPS
    //option.setLocationNotify(true);
    //可选，默认false，设置是否当GPS有效时按照1秒一次的频率输出GPS结果
    option.setIgnoreKillProcess(true);   //可选，默认true，定位SDK内部是一个Service，并放到独立
                                         //进程，设置是否在stop的时候杀死这个进程，默认不杀死
    option.setOpenGps(true);             //打开GPS
    //设置坐标类型，不设置的话，默认类型会有偏移，误差为700m
    option.setCoorType("bd09ll");        //设置坐标类型，百度地图SDK采用的是百度自有的地理坐标系
                                         //（bdll09）
    mLocationClient.setLocOption(option);
}
```

12.3.5 定位监听

```
public class MyLocationListener implements BDLocationListener {

    @Override
    public void onReceiveLocation(BDLocation location) {
        if (null != addressEdit) {
            addressEdit.setText(location.getAddrStr());
```

```
        }
        this.location = location;

    }

}
```

12.3.6 初始化地图 View

MapView 是用来显示地图的 View，BaiduMap 是 MapView 的操作类。

```
/**
 * 初始化地图
 */
protected void initMapView() {
    //地图初始化
    mMapView = (MapView) findViewById(R.id.bmapView);
    mBaiduMap = mMapView.getMap();
    //定位到上海
    //mBaiduMap.setMapStatus(MapStatusUpdateFactory.newLatLng(GEO_SHANGHAI));
    //设置地图比例尺，18级。比例尺为200m
    mBaiduMap.setMapStatus(MapStatusUpdateFactory.zoomTo(16));
    //开启定位图层
    mBaiduMap.setMyLocationEnabled(true);

}
```

12.3.7 显示位置信息

```
mBaiduMap.setOnMapClickListener(new BaiduMap.OnMapClickListener() {

            @Override
            public boolean onMapPoiClick(MapPoi arg0) {
                return false;
            }

            @Override
            public void onMapClick(LatLng arg0) {
                //System.out.println("projection = "+mBaiduMap.getProjection());


                MarkerOptions().position(arg0).icon( BitmapDescriptorFactory.fromResource
                    (R.drawable.icon_marka));

                Toast.makeText(context, arg0.toString(),Toast.LENGTH_SHORT).show();
                transLatLngToAddress(arg0);

            }
        });
```

12.3.8 获取当前屏幕的经纬度范围

```
//地图View加载完成的回调，mBaiduMap.getProjection()写在这里面，否则报空指针
    mBaiduMap.setOnMapLoadedCallback(new BaiduMap.OnMapLoadedCallback() {

            @Override
            public void onMapLoaded() {
                getLatLon();

            }
    });
```

1. 由屏幕坐标转换为对应的经纬度，计算屏幕的经纬度范围

```
protected void getLatLon() {
    //屏幕左上角的坐标点
    Point pointLeftTOP = new Point(0, 0);
    Projection project = mBaiduMap.getProjection();
    LatLng latlongLeftTOP = project.fromScreenLocation(pointLeftTOP);
    int width = ScreenUtil.getScreenWidth(context);
    int height = ScreenUtil.getScreenHeight(context);
    //屏幕右下角的坐标点
    Point pointRightBottom = new Point(width, height);
    //由坐标点得到经纬度
    LatLng latlongRightBottom = mBaiduMap.getProjection()
            .fromScreenLocation(pointRightBottom);
    String bounds = "latlongLeftTOP = " + latlongLeftTOP.toString()
            + ", latlongRightBottom = " + latlongRightBottom.toString();
    Toast.makeText(context, bounds, Toast.LENGTH_SHORT).show();
}
```

2. 用百度自带的方法获取屏幕经纬度范围

```
private void getScreenLatlon() {
    MapStatus status = mBaiduMap.getMapStatus();
    //得到地图中心对应屏幕坐标
    Point point = status.targetScreen;
    //得到地图中心经纬度
    LatLng latlon = status.target;
    LatLngBounds bouns = status.bound;
    LatLng southwest = bouns.southwest;
    LatLng northeast = bouns.northeast;

}
```

12.3.9 增加多个标注并监听

1. 增加标注

用 List 搜集标注，便于对标注处理。

```
BitmapDescriptor bitmap = BitmapDescriptorFactory.fromResource(R.drawable.icon_marka);
rivate void makrerOverlay(Rows bean) {
```

```java
        double lat = Double.valueOf(bean.getArea_lat());
        double lon = Double.valueOf(bean.getArea_lon());
        //定义Marker坐标点
        LatLng point = new LatLng(lat, lon);
        //构建Marker图标
        //构建MarkerOption，用于在地图上添加Marker
        MarkerOptions option = new MarkerOptions().position(point).icon(bitmap);
        //Marker出现的方式，从天上掉下
        option.animateType(MarkerAnimateType.drop);
        //在地图上添加Marker并显示
        Overlay overLay = mBaiduMap.addOverlay(option);
        Marker mMarkerA = (Marker) (overLay);
        map.put(mMarkerA, bean);

    }
```

2. 标注监听

```java
    @Override
    protected void initListener() {
        //点击泡泡
        mBaiduMap.setOnMarkerClickListener(new BaiduMap.OnMarkerClickListener() {
            @Override
            public boolean onMarkerClick(Marker marker) {
                clickMarker(marker);
                return true;
            }
        });
    }
    private void clickMarker(Marker marker) {
        Button button = new Button(getApplicationContext());
        button.setBackgroundResource(R.drawable.popup);
        OnInfoWindowClickListener listener = null;
        Set<Marker> set = map.keySet();
        for (Marker mark : set) {
            if (mark == marker) {
                final Rows data = map.get(marker);
                button.setText(data.getTitle());
                listener = new OnInfoWindowClickListener() {
                    @Override
                    public void onInfoWindowClick() {
                        mBaiduMap.hideInfoWindow();
                    }

                };
                break;
            }
        }
        mBaiduMap.showInfoWindow(mInfoWindow);
    }
```

12.4 关键知识点解析

12.4.1 在线更新

1. 获取应用的当前版本号

获取当前应用版本号,代码如下:

```
public static int getVersionCode(Activity activity) {
    PackageManager packageManager = activity.getPackageManager();
    PackageInfo packInfo = null;
    try {
        packInfo = packageManager.getPackageInfo(activity.getPackageName(),0);
    } catch (NameNotFoundException e) {
        e.printStackTrace();
    }
    return packInfo.versionCode;
}
```

2. 获取服务器版本号

本地版本获取成功后,再获取服务器版本号,判断是否需要更新,代码如下:

```
public static UpdataInfo getNewVersion() {
    UpdataInfo appInfo = null;
    try {
        URL url = new URL("http://");
        //开启一个连接
        HttpURLConnection connection = (HttpURLConnection) url.openConnection();

        connection.setConnectTimeout(5000);
        connection.setReadTimeout(5000);
        connection.setRequestMethod("GET");
        if(connection.getResponseCode()==200) {
            InputStream is = connection.getInputStream();
            InputStreamReader isReader = new InputStreamReader(is);
            String json = StreamUtil.streamToString(isReader);

            JSONObject jsonObject = new JSONObject(json);
            String versionName = jsonObject.getString("versionName");
            String versionDes = jsonObject.getString("versionDes");
            String versionCode = jsonObject.getString("versionCode");
            String DownloadUrl = jsonObject.getString("downloadUrl");

            appInfo = new UpdataInfo();
            appInfo.setVersion(versionCode);
            appInfo.setDescription(versionDes);
            appInfo.setUrl(DownloadUrl);
        }
```

```java
        } catch (Exception e) {
        }
        return appInfo;
    }
}
```

3. 下载文件

如果需要更新,则需要下载文件,代码如下:

```java
public class UpdateService extends Service {
    private NotificationManager nm;
    private Notification notification;
    private File tempFile = null;
    private boolean cancelUpdate = false;
    private MyHandler myHandler;
    private int download_precent = 0;
    private RemoteViews views;
    private int notificationId = 1234;

    @Override
    public IBinder onBind(Intent intent) {
        return null;
    }

    @SuppressWarnings("deprecation")
    @Override
    public void onStart(Intent intent, int startId) {
        super.onStart(intent, startId);
    }

    @SuppressWarnings("deprecation")
    @Override
    public int onStartCommand(Intent intent, int flags, int startId) {
        nm = (NotificationManager) getSystemService(NOTIFICATION_SERVICE);
        notification = new Notification();
        notification.icon = android.R.drawable.stat_sys_download;
        notification.tickerText = getString(R.string.app_name) + "更新";
        notification.when = System.currentTimeMillis();
        notification.defaults = Notification.DEFAULT_LIGHTS;
        //设置任务栏中下载进程显示的Views
        views = new RemoteViews(getPackageName(), R.layout.update);
        notification.contentView = views;

        PendingIntent contentIntent = PendingIntent.getActivity(this, 0,new Intent(), 0);
        notification.setLatestEventInfo(this, "", "", contentIntent);
        //将下载任务添加到任务栏中
        nm.notify(notificationId, notification);

        myHandler = new MyHandler(Looper.myLooper(), this);
```

```java
        //初始化下载任务内容Views
        Message message = myHandler.obtainMessage(3, 0);
        myHandler.sendMessage(message);
        //启动线程开始执行下载任务
        downFile(intent.getStringExtra("url"));
        return super.onStartCommand(intent, flags, startId);
    }

    @Override
    public void onDestroy() {
        super.onDestroy();
    }

    //下载更新文件
    private void downFile(final String url) {
        new Thread() {
            public void run() {
                try {
                    HttpClient client = new DefaultHttpClient();
                    //params[0]代表连接的URL
                    HttpGet get = new HttpGet(url);
                    HttpResponse response = client.execute(get);
                    HttpEntity entity = response.getEntity();
                    long length = entity.getContentLength();
                    InputStream is = entity.getContent();

                    File rootFile;
                    if (is != null) {
                        rootFile = new File("/sdcard/callCar");
                        if (!rootFile.exists() && !rootFile.isDirectory())
                            rootFile.mkdir();

                        tempFile = new File("/sdcard/callCar/"+ url.substring(url.lastIndexOf("/") + 1));
                        if (tempFile.exists())
                            tempFile.delete();
                        tempFile.createNewFile();

                        //已读出流作为参数创建一个带有缓冲的输出流
                        BufferedInputStream bis = new BufferedInputStream(is);

                        //创建一个新的写入流，将读取到的图像数据写入到文件中
                        FileOutputStream fos = new FileOutputStream(tempFile);
                        //以写入流作为参数创建一个带有缓冲的写入流
                        BufferedOutputStream bos = new BufferedOutputStream(fos);

                        int read;
```

```java
                    long count = 0;
                    int precent = 0;
                    byte[] buffer = new byte[1024];
                    while ((read = bis.read(buffer)) != -1 && !cancelUpdate) {
                        bos.write(buffer, 0, read);
                        count += read;
                        precent = (int) (((double) count / length) * 100);
                        //每下载完成5%就通知任务栏修改下载进度
                        if (precent - download_precent >= 5) {
                            download_precent = precent;
                            Message message = myHandler.obtainMessage(3,precent);
                            myHandler.sendMessage(message);
                        }
                    }
                    bos.flush();
                    bos.close();
                    fos.flush();
                    fos.close();
                    is.close();
                    bis.close();
                }

                if (!cancelUpdate) {
                    Message message = myHandler.obtainMessage(2, tempFile);
                    myHandler.sendMessage(message);
                } else {
                    tempFile.delete();
                }
            } catch (ClientProtocolException e) {
                Message message = myHandler.obtainMessage(4, "下载更新文件失败");
                myHandler.sendMessage(message);
            } catch (IOException e) {
                Message message = myHandler.obtainMessage(4, "下载更新文件失败");
                myHandler.sendMessage(message);
            } catch (Exception e) {
                Message message = myHandler.obtainMessage(4, "下载更新文件失败");
                myHandler.sendMessage(message);
            }
        }
    }.start();
}
```

4. 下载进度显示

在下载过程中显示下载进度，代码如下：

```java
/* 事件处理类 */
class MyHandler extends Handler {
    private Context context;
```

```java
        public MyHandler(Looper looper, Context c) {
            super(looper);
            this.context = c;
        }

        @Override
        public void handleMessage(Message msg) {
            super.handleMessage(msg);
            if (msg != null) {
                switch (msg.what) {
                    case 0:
                        Toast.makeText(context, msg.obj.toString(),Toast.LENGTH_SHORT).show();
                        break;
                    case 1:
                        break;
                    case 2:
                        //下载完成后清除所有下载信息，执行安装提示
                        download_precent = 0;
                        nm.cancel(notificationId);
                        Instanll((File) msg.obj, context);
                        //停止当前的服务
                        stopSelf();
                        break;
                    case 3:
                        //更新状态栏上的下载进度信息
                        views.setTextViewText(R.update_id.tvProcess, "已下载"+ download_precent + "%");
                        views.setProgressBar(R.update_id.pbDownload, 100,download_precent, false);
                        notification.contentView = views;
                        nm.notify(notificationId, notification);
                        break;
                    case 4:
                        nm.cancel(notificationId);
                        break;
                }
            }
        }
    }
}
```

5. 安装 apk

下载完成后，安装 apk，代码如下：

```java
    private void Install(File file, Context context) {
        Intent intent = new Intent(Intent.ACTION_VIEW);
        intent.setFlags(Intent.FLAG_ACTIVITY_NEW_TASK);
        intent.setAction(android.content.Intent.ACTION_VIEW);
        intent.setDataAndType(Uri.fromFile(file),"application/vnd.android.package-archive");
        context.startActivity(intent);
    }
```

12.4.2 Android 的四种定位方式

Android 定位一般有四种方法,即 GPS 定位、Wi-Fi 定位、基站定位和 AGPS 定位。下面进行详细讲解。

1. GPS 定位

GPS 定位需要 GPS 硬件支持,直接和卫星交互来获取当前经纬度信息,这种方式需要手机支持 GPS 模块。

优点:速度快、精度高、可在无网络情况下使用。

缺点:
- 比较耗电。
- 绝大部分用户默认不开启 GPS 模块。
- 从 GPS 模块启动到获取第一次定位数据,可能需要比较长的时间。
- 室内几乎无法使用。

使用位置服务功能,可以使用 LocationManager 位置服务管理器类。

要使用 Android 平台的 GPS 设备,需要添加如下权限:

```
<uses-permission> android:name= android.permission.ACCESS_FINE_LOCATION
</uses-permission>
```

2. Wi-Fi 定位

每一个无线 AP(路由器)都有一个全球唯一的 MAC 地址。Wi-Fi 定位是根据 Wi-Fi MAC 地址,通过收集到的该 Wi-Fi 热点的位置,将这些能够标示 AP 的数据发送到位置服务器,服务器检索出每一个 AP 的地理位置,并结合每个信号的强弱程度,计算出设备的地理位置获得经纬度坐标。

优点:受环境影响较小,只要有 Wi-Fi 的地方就可以使用。

缺点:
- 需要有 Wi-Fi,精度不准。
- 位置服务商要不断更新、补充自己的数据库,以保证数据的准确性。

从 Google 获取 Wi-Fi 定位新闻,代码如下:

```java
public static Location getWIFILocation(WifiInfo wifi) {
    if(wifi == null) {
        return null;
    }
    DefaultHttpClient client = new DefaultHttpClient();
    HttpPost post =new HttpPost( "http://www.google.com/loc/json" );
    JSONObject holder =new JSONObject();
    try{
        holder.put( "version" ,"1.1.0");
        holder.put( "host" , "maps.google.com" );
        JSONObject data;
        JSONArray array =new JSONArray();
        if(wifi.getMacAddress() != null && wifi.getMacAddress().trim().length() >0) {
            data =new JSONObject();
```

```
                data.put( "mac_address" , wifi.getMacAddress());
                data.put( "signal_strength" ,6);
                array.put(data);
            }
            holder.put("wifi_towers" ,array);

            StringEntity se = new StringEntity(holder.toString());
            post.setEntity(se);
            HttpResponse resp =client.execute(post);
            int state =resp.getStatusLine().getStatusCode();
            if(state == HttpStatus.SC_OK) {
                HttpEntity entity =resp.getEntity();
                if(entity != null) {
                    BufferedReader br = new BufferedReader(
                            newInputStreamReader(entity.getContent()));
                    StringBuffer sb = newStringBuffer();
                    String resute = ;
                    while((resute =br.readLine()) != null) {
                        sb.append(resute);
                    }
                    br.close();

                    data = new JSONObject(sb.toString());
                    data = (JSONObject)data.get( location );
                    Location loc = new Location(
                            android.location.LocationManager.NETWORK_PROVIDER);
                    loc.setLatitude((Double)data.get( latitude ));
                    loc.setLongitude((Double)data.get( longitude ));
                    loc.setAccuracy(Float.parseFloat(data.get( accuracy )
                            .toString()));
                    loc.setTime(System.currentTimeMillis());
                    return loc;
                }else {
                    return null;
                }
            }else {

                return null;
            }
        }catch (Exception e) {

            return null;
        }
    }
}
```

3. 基站定位

基站定位一般应用于手机用户。手机基站定位服务又叫作基于位置服务（Location Based

Service，LBS），它通过电信移动运营商的网络（如 GSM 网）获取移动终端用户的位置信息（经纬度坐标）。基站定位一般有两种，第一种是利用手机附近的三个基站进行三角定位，由于每个基站的位置是固定的，利用电磁波在这三个基站间中转所需要的时间来算出手机所在的坐标；第二种则是利用获取到的最近基站的信息，其中包括基站 ID、location area code、信号强度等，将信息发送到 Google 的位置服务，就能拿到当前所在的位置信息，误差一般在几十米到几百米之内。其中信号强度这个数据很重要。

优点：受环境影响的情况较少，只要有基站，不管在室内还是人烟稀少的地方都能用。

缺点：需要消耗流量，精度没有 GPS 那么准确，误差大概在十几米到几十米之间。

4．AGPS 定位

AGPS（Assisted GPS，网络辅助 GPS）使用设备的 GPS 芯片和移动电话网络来实现定位，利用网络来进行数据传输，以缩减 GPS 芯片获取卫星信号的延迟时间。

AGPS 广泛应用在移动设备中。

12.5　问题与讨论

1．如何根据所处位置进行 GPS 和基站定位优先切换？
2．对于百度定位偏差问题如何优化？
3．如何解决百度地图不显示地图的问题？
4．如何解决百度地图默认定位是北京的问题？